Acknowledgments

The authors acknowledge the support of the Project Nabimat funded by the Distretto ad Alta Tecnologia sui Sistemi Avanzati di Manifattura della Sicilia (DISAM district).

Contents

Chapter 1

Imperfect Uncertain Systems

1.1 Introduction

The main aim of this book is to introduce a new class of systems, namely the class of imperfect uncertain systems, and to propose new techniques to control their behavior. The idea comes from the simple observation that, even if the concept of uncertainty is well known, to characterize uncertain systems due to technological and real world imperfections implies a series of considerations aimed at understanding their behavior and the suitable control strategy to be adopted, when imperfections prevent us from obtaining a precise system.

Let us start introducing a rough distinction between two wide classes of systems: the class of systems that "does" work and the class of the system that "does not work". We intentionally use this informal language to identify those systems which, despite the presence of technological imperfections, are able to perform the designed task. In this wide classification, we intend mainly to distinguish between systems, which reach stable working conditions and others that do not reach them.

In our mind, perfection is intrinsic. The standard guidelines followed by engineers in the design of physical devices are based on the assumption that perfect systems can be realized. Improved technology leads us to work with an increasing level of precision in realizing the systems, working with quasi-ideal systems. In these cases, the lack of uncertainty ensures the possibility to avoid feedback control, choosing open loop control schemes.

Ideal systems are the goal of the next technology. We can state that this is the age of perfection! What really comes out is the recurrent concept of perfection from a philosophical point of view — it is an objective that must be achieved, and in a parallel way the technological perfection has

followed the philosophical concept. Today, the age of technological perfection is a day-by-day challenge that must be pursued. Examples are in
electronics and in micro-machinery mechanical technology. Nanotechnology
has a fundamental task: it operates in a small and small size dimension,
by achieving with high precision what the chemical reaction in solid state
physical system achieves in a controlled way.

Moreover, what appears in the ideas of the twentieth century from a
philosophical point of view is the trend that the concept of perfection has
changed thanks to the growing interest for a non-idealist idea of perfection.
This is due to various reasons: first of all, the review of science, the new
concepts of statistical physics, quantum physics and the Gödel's theorem
have contributed to this new perspective. In addition, history, anthropo-
logy, social sciences have led to perfection to be reconsidered. From another
point of view, the world is moving in another direction to push technology
towards perfection. And it is impossible to attain. We know that uncertainty always exists, and, what is more, uncertainty and imperfection go
together.

In this book, after having distinguished the concept of uncertain and
imperfect systems, the development of a strategy to obtain models for the
systems of this class is faced. The concepts related to this class of systems will be highlighted. The possibility to control them, in order to make
them working, or, better, to make them working in stable operating conditions will be discussed. In order to achieve this target, the new concept of
additional *hidden*, or *imperfect*, dynamics will be introduced.

In this scenario, the central role of noise and broadband signals in for-
mulating the control strategy will be widely discussed. And the idea to
use a feedback controller based on these types of signals will be proposed.
Therefore, a general theory to identify and to control this class of systems
will be outlined.

The book will illustrate the proposed strategy for the control of more
structures of electromechanical systems. Moreover, it will be emphasized
that, even if the introduced theory has been conceived to design control systems, its main concept is very useful to derive a new class of systems whose
behavior is determined by the role that uncertainty, noise and imperfection
play.

1.2   Examples

The ideas of considering imperfect uncertain systems arouse from observations and experiments. In this section, we will provide some interesting examples of imperfect uncertain systems which allow us to clarify the main concepts introduced in the previous section.

1.2.1   *Friction in a sliding wheel*

The existence of imperfections in mechanical systems is something which can be somehow easily imagined. Let us consider the schematic representation of a wheel sliding over a surface reported in Fig. 1.1. The upper panel of Fig. 1.1 represents the ideal case, in which friction is minimized assuming that the wheel section is perfectly circular and that the surface over which the wheel slides is regular. In this case, the system *works* exactly as from ideality, and the wheel indefinitely slides over the surface. The lower panel of Fig. 1.1 reports the opposite case, both the wheel and the surface are far from ideality and their real physical properties prevent the desired sliding motion. However, in reality, systems like these continue to *work* despite imperfections, as reported in the middle panel of Fig. 1.1. Moreover irregularities, which are intrinsic and unavoidable in the construction process of both wheels and surfaces, do not affect the working principle of the wheel. The fact that real systems are imperfect is therefore strictly related to the

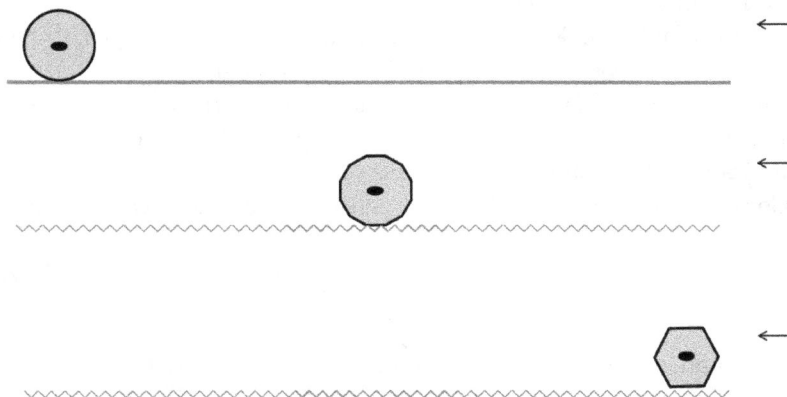

Fig. 1.1   Effect of friction on a sliding wheel: (a) ideal case, no imperfections, (b) real case, imperfections do not affect the correct behavior, (c) real case, high level of imperfections prevent sliding of the wheel.

imperfection of the construction processes, i.e. imperfections are introduced passing from the ideality of the design to the reality of the physical
realization.

1.2.2   *Nonidealities in electronic devices*

Electronic integrated devices are realized through fabrication process of
high complexity in order to ensure that the designed behavior is actually
implemented. However, during the integration of the device, unwanted
effects due to parasitic capacitances and inductances arise. These parasitic dynamics may sensibly shift the real behavior far from the ideality.
Let us consider the case of analog multipliers based on the four-quadrant
cell, which are known to have nonideal memory effects [Baranyi and Chua
(1982)].

The input-output characteristic of an ideal analog multiplier is

$$V_{out} = V_x(t)V_y(t) \tag{1.1}$$

with $V_x(t)$ and $V_y(t)$ being the two input voltages. However, a dynamic
model considering the memory effect should involve time-derivatives of the
inputs [Buscarino *et al.* (2016a)] so that the output of the real analog multiplier is given by

$$V_{out} = K(V_x(t)V_y(t) - T_A\dot{V}_x(t)V_y(t) - T_BV_x(t)\dot{V}_y(t)) \tag{1.2}$$

where K, T_A, and T_B are model parameters characteristic of the multiplier.
The values of the model parameters, which spoil the correct multiplication
of the two input signals, have to be estimated and are revealed to be de-
pendent on the frequency spectrum of the input signals.

The possibility to estimate the imperfections and their effects on the dynamical property of the device can be actually exploited to obtain complex
dynamics with few simple components.

Consider the following third-order autonomous dynamical system:

$$\begin{aligned}
\dot{x} &= y - ax \\
\dot{y} &= \gamma x - Rx^3 + \Gamma z - by \\
\dot{z} &= \alpha - \beta z^2 - x^2
\end{aligned} \tag{1.3}$$

where a and b are positive dissipation rates and γ, R, α, β and Γ are
system parameters. Here, we fix $a = b = 0.1$, $R = 1.09$, $\Gamma = 0.7$, $\alpha = 1$ and
$\beta = 0.001$ and obtain the bifurcation diagram shown with the blue curve
in Fig. 1.2. For the range of γ considered, there is always a period-1 limit
cycle with constant amplitude.

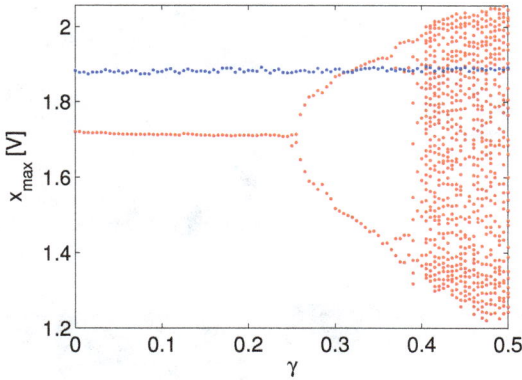

Fig. 1.2   Experimental bifurcation diagram with respect to the parameter γ shown in the local maxima of the state variable x for the circuit implemented using analog multipliers (red) and replacing analog multipliers with a diode-based device implementing the PWL approximation of $F = x^3$ (blue).

Now consider a slight modification of Eqs. (1.3) including a further nonlinearity $F(x, \dot{x})$ given by

$$\begin{aligned}
\dot{x} &= y - ax \\
\dot{y} &= \gamma x - RF(x, \dot{x}) + \Gamma z - by \\
\dot{z} &= \alpha - \beta z^2 - x^2
\end{aligned} \tag{1.4}$$

where the form of $F(x, \dot{x})$ sensibly affects the emergent behavior of the system in Eqs. (1.3). In particular, the first-order derivative of x gives the possibility of chaotic dynamics, when $F(x, \dot{x}) = x^3 - \delta\dot{x}x^2$, where δ is a further system parameter. For $\delta = 0.16$, the system in Eqs. (1.4) shows a bifurcation diagram with respect to γ that includes a cascade of period doublings towards a chaotic window shown with the red curve in Fig. 1.2. In Fig. 1.3, three different projections of the attractor obtained for $\gamma = 0.45$ are shown.

The onset of a chaotic behavior is, therefore, strictly linked to the modification of the nonlinear function F which is now the conjunction of two terms: a static nonlinearity x^3, and a product nonlinearity $\dot{x}x^2$ involving a square and a derivative operation. This modification corresponds to the explicit introduction of nonideal terms for two cascaded analog multipliers realizing the cubic operation. In this case, the existence of imperfections in real analog multipliers lead to an added-value in the design phase. Whereas the circuit implementation of $F(x, \dot{x})$ involves the use of differentiators, which has intrinsic instability at high frequency, using just two imperfect

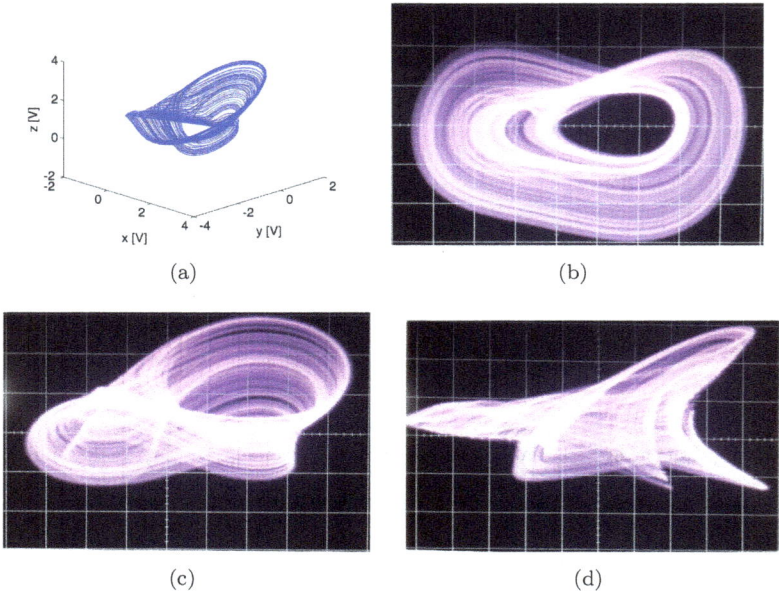

Fig. 1.3 (a) 3D plot and (b) oscilloscope traces of the attractor on the xy-plane, (c) xz-plane, and (d) yz-plane from the circuit with $\gamma = 0.45$. Scale of oscilloscope traces is 500mV/div.

analog multipliers ensures the introduction of the needed nonlinear terms by suitably designing the frequency range of the circuit.

1.2.3 *Oscillations in a single transistor circuit*

The oscillator we are reporting here is a further example regarding the relevance of imperfections that can be conceived with electronic components. The scheme of the circuit is reported in Fig. 1.4 and consists of an RC circuit powered by a DC voltage and a single NPN 2N2222 transistor, where only the emitter and the collector are used, and the base is floating. This oscillator is based on the Pearson and Anson effect, discovered in 1922 observing an oscillation when a powered RC circuit was connected with a neon bulb in parallel to the capacitor. The equivalent effect is obtained by using the tunnel effect in BJT for which Leo Esaki had the Nobel Prize in 1973. In our scheme, the transistor is configured as a negative differential resistor (NDR) [Sprott (2010)].

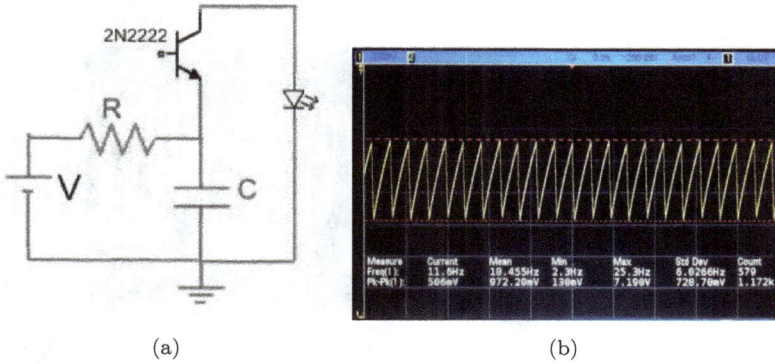

Fig. 1.4 Oscillations due to imperfections in a single transistor circuit: (a) implementation of the circuit (component values: $R = 1.5\text{k}\Omega$, $C = 220\mu\text{F}$, 2N2222 BJT, $V_s = 10V$); (b) temporal trend of the oscillation.

The voltage across the capacitor increases with a time-constant RC until the breakdown voltage of the BJT is reached and the emitter and the collector are then short circuited. The capacitor undergoes a fast discharge and the process cyclically continues. Even if the frequency of the oscillator is mainly linked to the RC time-constant, it also depends on the discharging time of the capacitor. But, most of all, it relies on the exact value of the breakdown voltage which is related to the imperfections of the single device.

1.2.4 *Chaotic behavior of operational amplifiers*

We discuss here a last example regarding the role of imperfections in the generation of complex dynamical behavior. Let us consider the configuration reported in Fig. 1.5(a), in which a single operational amplifiers $LF357N$ is connected in a negative feedback scheme. Despite the absence of external inputs, excepting the power supply, and memory elements, a chaotic behavior can be observed at the output terminal of the amplifier [Yim *et al.* (2004); Buscarino *et al.* (2017)]. Here we can fix $R_1 = 20\Omega$ and $R_2 = 0\Omega$, a chaotic behavior is observed when the dual power supply of the operational amplifiers is unbalanced as $V_+ = 7V$ and $V_- = -4.3V$ as shown in Fig. 1.5(b).

However, the exact values of power supply and R_1 leading to the onset of chaos strongly depend on the specific device used. This is due to the fact that chaos appears as a consequence of the parasitic capacitors which

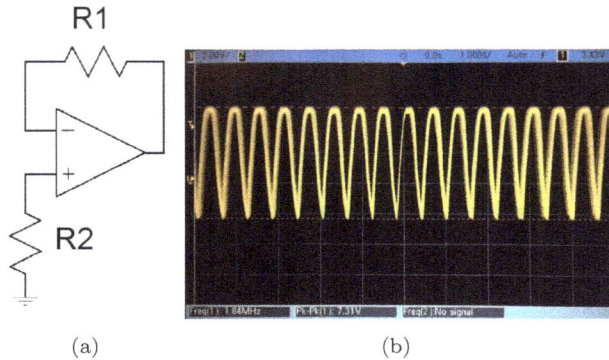

Fig. 1.5   Chaotic behavior in a operational amplifier: (a) implementation of the circuit, (b) temporal trend of the oscillator.

are inside the operational amplifier, whose values depend on the fabrication process of the single device. Therefore, the presence of imperfections in the $LF357N$ is emphasized in this very simple scheme.

1.2.5   *Chaotic actuation of a microrobot*

The useful role of imperfections is also enlightened by a further example. Let us consider the microrobot reported in Fig. 1.6 (a complete description of the microrobot and the related control system is reported in [Buscarino *et al.* (2007)]). It is constituted by a passive body and two active piezoelectric legs which are actuated through a specific locomotion scheme. Using a purely periodic locomotion pattern provides the robot a correct locomotion scheme which, however, works only when the surface over which the robot moves opposes a sufficient, but not large, friction. Imperfections over the sliding surface are in fact essential for robot motion. If friction is too low,

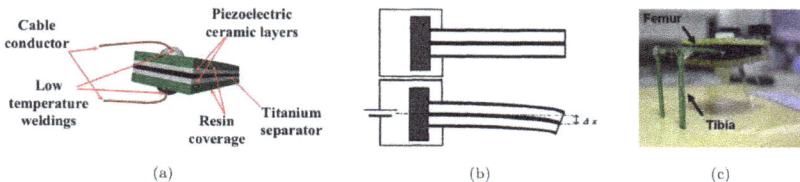

Fig. 1.6   (a) Piezoelectric scheme. (b) A piezoelectric actuator. (c) The microrobot structure.

legs will slide over the surface without inducing the motion of the passive body and, on the opposite, if friction is too high, the robot remains stacked since its motion is not able to overcome imperfections.

Including an irregular locomotion pattern, i.e. a locomotion scheme in which legs are not actuated periodically with a given fixed frequency, leads to a non-trivial scenario. Considering a locomotion pattern based on a time-varying frequency of actuation in which the frequency changes according to chaotic dynamics, the capabilities of the microrobot in overcoming surface imperfections are sensibly improved. This effect is visible in Fig. 1.7. To graphically show the improvements obtained using chaotically modulated frequency signal, we equipped our robot with a LED that lights up when the robot is actuated. We then use a camera with a long exposure time in order to take pictures with tracks of the robot trajectory. Improvements are visible simply comparing the length of the red lines in the Fig. 1.7.

This is a further confirmation that the interplay of different sources of imperfection eventually leads to an increased level of coordination and organization.

Fig. 1.7   Performances of the microrobot on a scratched wooden surface. (Left) The constant frequency case, and (right) the chaotically modulated frequency case.

1.3   Understanding Imperfect Uncertain Systems

The study of imperfect uncertain systems must tackle physical systems like those described in the previous section, i.e. systems working not only in the presence of but thanks to the existence of imperfections. Inside this technical concept, there are some intrinsic questions that arise. Indeed, the problem we are approaching can be seen in the larger context of the opponency between determinism and indeterminism. In these cases, in fact, the occurrence in the real life of the desired behavior seems to be not due to the sufficient conditions for which they should occur, as the classical determinism states.

1.3.1   *Stochastic vs. deterministic systems*

The concept of determinism has been developed in the 18th century, under the formulation that each event is caused by a precise action. This leads to conceive the idea that the future can be foreseen. Indeed, this formulation is not complete, because the causality relation offers more interpretations. In fact we cannot exclude that the events are caused in a non-deterministic way, where the cause does not determine the effect, but it increases the probability that it may occur. This leads to the introduction of stochastic systems.

There is a universal agreement both referring to the classical physics and the quantum mechanical physics that the non-determinism is related to measurements, and, therefore, to precision and to the degree of their approximation. The main idea described in this book is to consider in the proposed study the guidelines to examine the real systems in relation to the concept of imperfection and uncertainty.

Indeed, in the suggested definition, the qualitative behavior of the system and the possible causes that can arise to change its behavior are well known. We are conscious of the parametric uncertainty of the system. As a consequence, the state equations of the system are known and, in any case, a mathematical model of the system can be derived. The key point is to understand in which way imperfections work and how they positively affect the system behavior to make it *working*.

1.3.2   *Imperfections and uncertainties*

At this point, it is important to define the concept of imperfection. A formal definition is difficult, so that we must rely on a tautology: imperfections are all the ingredients that make the system not perfect. However, we can outline the following features that characterize imperfect systems:

(1) they are dynamical systems;
(2) they present a hidden parasitic dynamics;
(3) they contain threshold components;
(4) some parameters are suddenly changed by external stochastic noise;
(5) some parameters depend on the working conditions;

(6) an intrinsic network of switching effects leads the system from one behavioral condition to another one;

(7) silent dynamics characterizes them. This is the key point of imperfect
systems. We call this dynamics *hidden* or *imperfect dynamics* that will
help to control the overall behavior system;

(8) intermittent behavior characterizes the dynamics of imperfect systems.

Even if we deal with continuous-time imperfect systems, the main concept of imperfect systems is quite general. Until now, a general idea of the
system we deal with has been given; the following parts of the book will report various cases of imperfect systems and will introduce the mathematical
items to describe them.

In the next part of this section, attention will be paid to the concept
of uncertainty. The uncertainty in our framework is different from imperfection. Therefore, the two concepts must be distinguished. It is assumed
that uncertainty is due to the imponderable variability of the parameters
in the system. It may be caused by behavioral conditions, these are related
to time-varying physical parameters of the system and in any case we do
not approach this aspect by using stochastic models. It is postulated that
uncertainty exists!

Of course, in the ideal case we do not have either uncertainty nor imperfections and we have to consider only an ideal perfect system. And, even
if the effort of the system designer is to achieve the goal to have certain
and perfect systems, it is not generally an accomplishable task. Indeed,
uncertainty and nonlinearities in the model do intrinsically exist but good
approximations can be performed and in these cases the classical approaches of control theory, robust or not, lead to good control performance
[Slotine and Li (1991); Bhattacharyya *et al.* (1995)].

Summarizing the previous concept, in order to deeply understand the
working mechanism of imperfect uncertain systems, we have to know and
identify the equations of both the main systems and that of the hidden
dynamics.

This central goal must be clear: this means that the controlled behavior
of the system must be fixed and the limitation of the control of imperfect
uncertain systems must be very clear to the user and to the control engineer.
This implies to face the problems of the control of this class of systems given
their limits and establishing the performance limits that these systems can
reach. Moreover, new paradigms of control schemes must be adopted to
get a satisfactory trade-off between behavior and performance.

Even if some ideas will arise in dealing with imperfect uncertain control systems, some essential facts must be taken into account. First of all, to guarantee the correct behavior, and, second, to achieve suitable control performance. Furthermore, some other constraints may have to be taken into account, like energy consumptions, the limitation of mechanical stress and so on. Another critical point will be the design of the sensor-actuators systems, since their implementation must avoid unnecessary modifications of the imperfect system and to be as less invasive as possible.

1.4   Main Ideas to Control Imperfect Uncertain Systems

To face a new control problem, first of all, a global view of it must be realized. This means that in order to understand how to tackle the control problem, a wide range of scenarios must be taken into account and both the possible mathematical models and the physical knowledge of many examples of systems must be evaluated leading to an accurate evaluation of the control strategy also by proposing new ideas.

Even if the standard control strategies have to be the main guidelines in facing the problem, new ideas and new control paradigm must be established to simplify the problem and to achieve suitable results.

Indeed in dealing with the considered systems, we are going to face a crisis. Many concepts that we have in our background fail. It means that the control of an imperfect uncertain system implies a critical evaluation of well established control theories. This is clear when dealing with the real evaluation of imperfect uncertain systems.

First of all, new attempts to model the class of imperfect uncertain systems must be taken into account. Classical mathematical models help us to understand and to design preliminary models, but, for this class of systems, it is not sufficient to have the help of equations and physical relationships only to model them. Doing this involves both the realization of particular equipments and also new terminology and techniques that sensibly differ from that usually adopted.

This last consideration leads us to a bifurcation in the paradigm we usually adopt and the development of new paradigms to face the problem. This route has been followed in order to develop the main guidelines to approach imperfect uncertain systems.

Indeed, the following items must be included in the new paradigm:

(1) the perception of hidden dynamics. This is the key point. This aspect can involve the definition of its modeling strategies. Indeed, the accuracy of the model is not a crucial point. Instead, when we discuss about modeling of hidden dynamics, we intend to derive an approximate model, or a qualitative model, that reproduces the behavior of the real dynamics;

(2) a reasonable knowledge of the system is needed to understand how hidden dynamics can be stimulated;

(3) the understanding of how the dynamics of the system has to be controlled paying attention to the interaction of its various parts in order to establish the relationship between the systems and the hidden dynamics;

(4) the derivation of the strategy to control the system. In general this concept is well known, in our discussion we remark that the control law is established by the hidden dynamics. Therefore our task is to generate the control strategy in order to stimulate and to control the hidden dynamics. Our task is to develop the control strategy in order to guarantee that the hidden dynamics does work to control the main system. The problem is in this sense reversed. A new paradigm arises;

(5) a qualitative model of the system. This can be done by analogies and comparisons with well known models of systems that work like the considered one or by establishing new models based on cognitive approaches;

(6) the possibility of using on-off control laws where the compensator generates sequences of signals to stimulate or not the hidden dynamics.

The control of this new class of systems reflects the main items of the classical control schemes but taking into account the following items:

- the performance of the system is limited;
- the system behavior should be robust both in terms of stability and performance;
- the energy management and optimization of the whole system must be taken into account.

In order to achieve the previous items, a suitable set of measurements must be taken into account. Moreover, only low energy control actions must be performed, as the real multi-loop control action is performed by the hidden dynamics.

1.5   Case Studies

In the following sections, we will introduce and describe paradigmatic examples of imperfect uncertain systems. All of them are based on a simple building block, consisting of a rotating coil, actuated by the interplay between the magnetic field imposed by a magnet and the electric field induced by a current flowing in the coil.

This simple device encompasses all the ingredients to build an imperfect uncertain system, since its realization process cannot be serialized. Therefore, each coil, each magnet associated to the coil, and each trail over which the coil is allocated will present unavoidable differences. Furthermore, coupling many of these simple blocks together, a complex structure can be easily realized, whose global behavior is strongly nonlinear and deeply influenced by the imperfections of each building block.

The case studies considered in this book range from low to high-scale systems of interacting rotating coils. In particular, we will focus on two electromechanical structures, such as those reported in Fig. 1.8. They consist of a set of trails hosting N slots.

These structures and their behavior will be characterized in the following sections. Furthermore, the suitable control actions to elicit the hidden dynamics and increase the performance of the systems will be described.

Beyond the mentioned electromechanical structures, we will focus also on two devices built over the same working principle of the rotating coils. The first experimental setup is reported in Fig. 1.9(a) and it is realized by two rotating coils actuating a flexible beam with a magnetic tip. The complex behavior of the flexible beam, which includes a cascade of bifurcations toward chaotic motion, will be described in terms of imperfect uncertain systems. Moreover, the motion of a single rotating coil, as that reported in Fig. 1.9(b), is able to show chaotic motion which will be also characterized in the following.

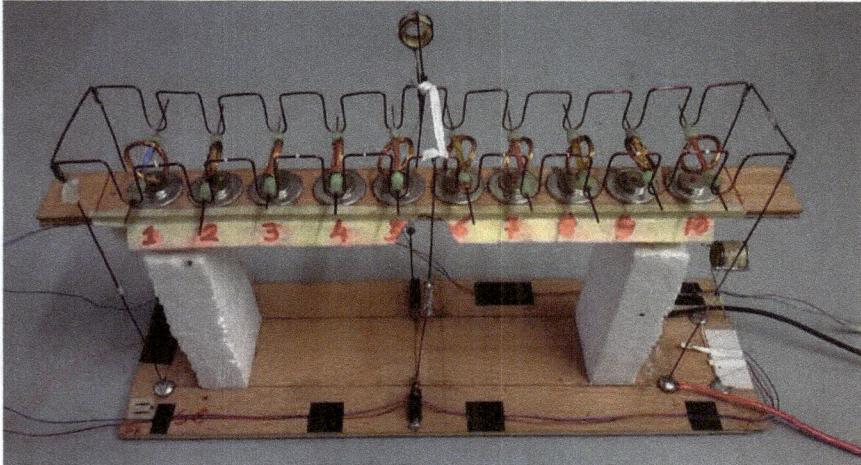

(a)

(b)

Fig. 1.8 Examples of imperfect uncertain systems by coupled rotating coils: (a) rectangular structure with $N = 10$ slots, (b) circular structure with $N = 16$ slots.

(a)

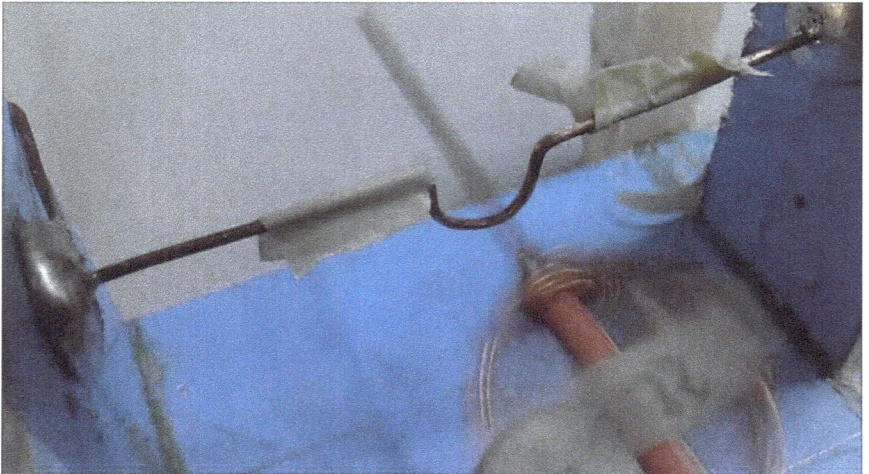

(b)

Fig. 1.9   Two other examples of imperfect uncertain systems: (a) two coil-magnet system with a flexible beam; (b) single coil-magnet system.

Chapter 2

Modeling and Control Strategy

2.1 General Remarks

Before describing the mathematical model of the considered systems the following general definitions and remarks are given.

An imperfect uncertain system is a system for which the presence of imperfections, nonidealities and uncertainties has to be explicitly taken into account in the model so that it can represent the real behavior observed in the physical system.

Let us consider, for example, the coils coupled through one of the flexible structures shown in Fig. 1.8. Ideally, the coils must be realized by using symmetric copper windings, must have exactly the same geometry and terminals which, under ideal conditions, must be 50% isolated and 50% conductive. The magnetic flux should also be exactly symmetric with respect to the coil position. Moreover, the electromechanical structure hosting the coils should have the ideal characteristics of a brush-pivot mechanical system and, so, both sides of each coil (in particular, the geometry of their realization), complain with the requirements of such ideal system.

However, this is not the real case. In fact, in dynamical conditions the pivot runs inside the brush keeping its rotation while climbing over sides of the slots and eventually falling down. If the slots are not symmetric and geometrically perfect, then the coil falls down differently from one side to the other. Therefore, in the model the ideal conditions have to be paired with a dynamics accounting for the onset of oscillations generated by the impacts of the coils sliding along their mechanical support and then collapsing into it.

The coil motion inside the slots where they are located is, in fact, due to the geometrical imperfections of the electromechanical structure and it is the reason for which horizontal and vertical vibrations arise in the structure.

2.2   Mathematical Model of an Imperfect Uncertain System

From a mathematical point of view, an imperfect uncertain system is a system described by a set of interacting state space equations, some of which model the imperfect dynamics. In particular, we consider continuous-time models, so that the model equations can be written as:

$$\dot{x} = f(x, \tilde{x}, u, w, p, t)$$
$$\dot{\tilde{x}} = \tilde{f}(x, \tilde{x}, u, w, p, t)$$

(2.1)

where:

- $x \in \mathbb{R}^n$ is the state vector of the system under ideal conditions;
- $\tilde{x} \in \mathbb{R}^k$ is the state vector of the imperfect dynamics;
- $u \in \mathbb{R}^m$ is the vector of the exogenous control signals;
- $w \in \mathbb{R}^q$ is the vector of exogenous not controllable signals;
- $p \in \mathbb{R}^g$ is the vector of the system parameters;

and $f : \mathbb{R}^n \times \mathbb{R}^k \times \mathbb{R}^m \times \mathbb{R}^q \times \mathbb{R}^g \times \mathbb{R} \to \mathbb{R}^n$, and $\tilde{f} : \mathbb{R}^n \times \mathbb{R}^k \times \mathbb{R}^m \times \mathbb{R}^q \times \mathbb{R}^g \times \mathbb{R} \to \mathbb{R}^n$ are nonlinear functions.

The block scheme representation of system (2.1) is illustrated in Fig. 2.1.

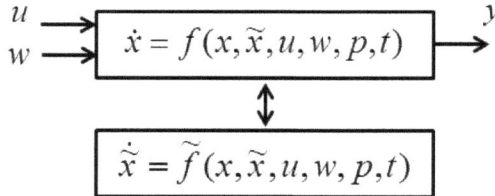

Fig. 2.1   Block scheme representation of system (2.1).

The imperfect uncertain systems have the following peculiarities: the parameter uncertainty may lead the system to not work; the imperfect dynamics, if it is excited, can stimulate the system to achieve the suitable parameter conditions that allow the emergence of its proper behavior. Therefore, from two factors usually considered as undesired (that is, the presence of uncertainties and imperfections), an adaptive strategy may be attained that, on the contrary, exploits these two factors in a positive way favoring the system working condition.

2.3 Vibrational Control

In this section, some insights on classical vibrational control strategies will be outlined. Furthermore, we will introduce the concept of closed-loop vibrational control in order to emphasize the main differences with the strategy proposed for imperfect uncertain systems.

2.3.1 *Classical vibrational control*

The main principles of vibrational control have been introduced by S.M. Meerkov in 1973 [Meerokov (1973)]. It consists essentially of an exogenous vibration realized in the system with a zero mean signal, and without modifying its structure. The principal difference with traditional feedback and feedforward control schemes is that vibrational control does work without measurements, and it is able to ensure the stability of the system, suitable performance, and the rejection of disturbances.

Let us consider a system whose dynamics is expressed by the following equation:

$$\dot{x} = f(x, t) \tag{2.2}$$

where $x \in \mathbb{R}^n$ is the state vector and $f : \mathbb{R}^n \times \mathbb{R} \to \mathbb{R}^n$. The vibrational control introduces the following modification in the model of the system:

$$\dot{\tilde{x}} = \tilde{f}(\tilde{x}, u, t) \tag{2.3}$$

where $\tilde{x} \in \mathbb{R}^n$ is the state vector of the controlled system, u is the vibrational control signal, and we assume that $\tilde{f}(\tilde{x}, t) = f(x, t)$ if $u = 0$. The main task is to choose $u(t)$ such that the required stability performance can be verified.

The Meerkov's vibrational control can be easily explained by using the paradigmatic example of the inverse pendulum. The pendulum represented in Fig. 2.2(a) is on an unstable equilibrium point, i.e. a small perturbation of the state tends to move it away from the equilibrium. If we apply a torque $M(t)$ at the pivot, of a high frequency, zero-mean periodic signal, as in Fig. 2.2(b), the original equilibrium can be stabilized.

The equation of an inverse pendulum can be written, neglecting the viscous effect, as:

$$ml^2\ddot{\theta} = mgl\sin\theta \tag{2.4}$$

where m is the pendulum mass, l the pendulum length, g the gravity, and θ the angular position of the pendulum with respect to the direction orthogonal to the plane of the pivot. By linearizing the system, it is easy

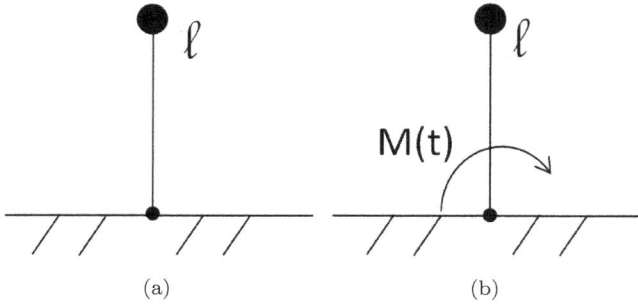

Fig. 2.2   Example of vibrational control applied to an inverse pendulum: (a) schematic representation of the uncontrolled system at the unstable equilibrium point, (b) application of a torque $M(t)$ at the pivot in order to stabilize the equilibrium point.

to check that the state $\theta = 0$ is unstable. The equation of the vibrational controlled system is given by:

$$ml^2\ddot{\theta} = mgl\sin\theta + M(t) \tag{2.5}$$

where $M(t)$ is the torque.

This example can be easily realized through a simple experimental setup in which a DC motor and a function generator are used. In Fig. 2.3(a) an inverse pendulum is made by a plastic bar attached to the rotor of a DC motor, implementing the actuated pivot. The motor, in fact, provides the small torque $M(t)$ on the basis of the signal used as power supply. Practically, we can control $M(t)$ by choosing the signal $V_a(t)$ which provides the power supply to the motor. The picture reported in Fig. 2.3(b) shows the effectiveness of this approach when choosing $V_a(t)$ as a square wave signal with zero mean value, amplitude of ± 3V and frequency of 4Hz. The picture has been obtained by lowering the shutter speed of the camera, so that the narrow motion around the unstable equilibrium can be appreciated.

In order to establish exactly the features of the vibrational control signal, the linearized equation of system (2.5) can be used and the averaging principle can be adopted considering the signal $M(t)$ as a time-varying coefficient in a time-invariant system. The amplitude and the frequency of the signal $M(t)$ can thus be derived so that the equilibrium point is stable. Only recently [Berg and Wickramasinghe (2015)], a technique for vibrational control without averaging has been proposed.

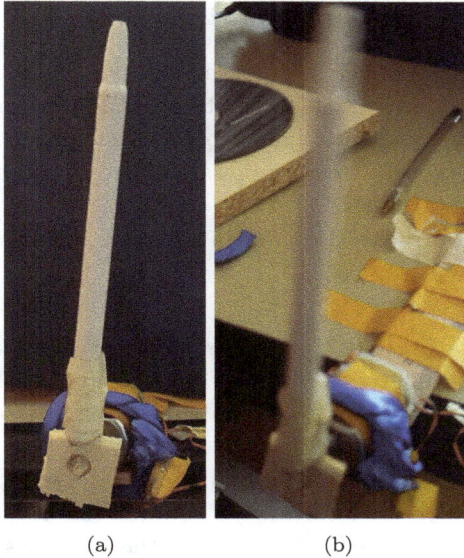

Fig. 2.3 Experimental realization of a vibrational control inverse pendulum: (a) experimental setup consisting of a plastic bar and a DC motor, (b) low shutter speed picture of the vibrational controlled system.

2.3.2 *Closed-loop vibrational control*

In 1993, Meerkov, Kabamba and Poh [Meerokov *et al.* (1993)] proposed a new control technique called closed-loop vibrational control. It has been applied to the suppression of vibrations in helicopter dynamics. The control signal adopted is periodic and respects the constraint of the classical vibrational control. Its amplitude, however, varies on the basis of the system output. The advantage of this method is that the classical feedback theory can be used to fix the closed-loop performance.

In order to formalize this approach, let us consider the class of systems described by the following state-space equations:

$$\begin{aligned}\dot{\mathbf{x}} &= \mathbf{A}\mathbf{x}(t) + \mathbf{B}u(t) \cdot f\left(\tfrac{t}{\varepsilon}\right) \\ \mathbf{y} &= \mathbf{C}\mathbf{x}\end{aligned} \tag{2.6}$$

where $0 < \varepsilon \leq 1$, $f(t) = f(t + T)$, with $T \neq 0$ and $\frac{1}{T}\int_0^T f(\tau)d\tau = 0$. In Eqs. (2.6), $u(t)$ is the control input whose amplitude is modulated by the open-loop vibrational control signal $f\left(\frac{t}{\varepsilon}\right)$.

2.3.3   *General remarks on vibrational control*

Vibrational control must not be confused with vibration control [Thomsen (2003)]. The latter, in fact, is devoted to attenuate or eliminate undesired vibration controlling the performance of the system in a either passive or active way. An approach based on vibrational control can be also suitable to attain a vibration control.

As concerns the strategy proposed in this book, the following items must be remarked:

- a closed-loop vibrational control is used, but it is not based on the evaluation of the output of the system to modulate the control signal. The output of the system is used only to switch on or off the controller;
- vibration control is not performed;
- our control actions consist in stimulating vibrations that act as active signals in order to drive the system towards the *working* condition;
- once the system is correctly working, the control signal intensity is lowered in order to drive the system towards a *regularized* behavior.

2.4   A Control Strategy for Uncertain Imperfect Systems

On the basis of the previous considerations, the natural question that arises is: how to stimulate the dynamics due to the imperfections?

The scheme of Fig. 2.4 shows the control principle for an imperfect uncertain system.

Excitatory signals in a closed-loop scheme are used to stimulate the imperfect dynamics that exerts the real control action on the system.

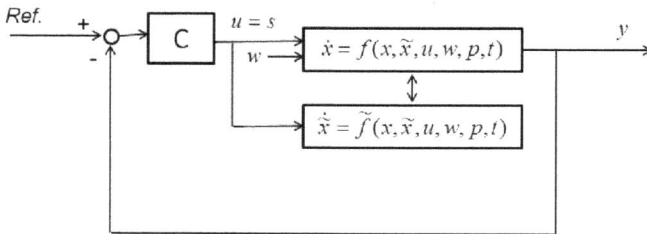

Fig. 2.4   Control scheme for an imperfect uncertain system.

2.4.1 *Vibrational control signals: periodic and chaotic actuation*

A further relevant point regards the nature of the vibrational control signals used in the control of imperfect uncertain systems. Following a classical approach [Sun (2006)], we will explore the possibility to use periodic control signals which are able to elicit a given mode of the system. In this case to choose the suitable control frequency, a preliminary analysis on the frequency response of the system must be carried out. In particular, the frequency response between a given control input u_i and the output of the system has to be estimated for the physical system. Furthermore, this analysis must include the evaluation of the spectral coherence, in order to estimate the causality relationship between the given input and the considered output.

The choice of a periodic control signal may be useful in controlling imperfect uncertain systems, even if following this strategy a fundamental aspect of this class of systems tends to be neglected. Imperfect uncertain systems, in fact, are characterized by several sources of nonideality which eventually lead to the coexistence of different characteristic modes which govern different parts of the considered physical system. To account for this peculiar behavior, a broad spectrum control signal may play a crucial, and more effective, role in controlling the system to the working conditions.

A broad spectrum signal can be generated by means of stochastic or deterministic processes. A Gaussian noise is indeed a broad spectrum signal and may be used to excite all the modes of the imperfect uncertain system. However, using noise to define a control law has the drawback that it is not possible to obtain the very same stochastic signal in two different experimental sessions.

On the other hand, deterministic processes can also provide broad spectrum signals, such as chaotic oscillations. A chaotic signal is characterized by a wide spectrum and an intrinsic irregular behavior with a high sensitivity to initial conditions. Furthermore, since it is generated by a deterministic system, chaotic behavior can be controlled, thus allowing to fix a given control law chosen according to the response of the imperfect uncertain system [Bucolo *et al.* (2002); Gacia-Ojalvo and Sancho (1999)].

In the following, the use of chaotic control signals will be investigated, since their rich frequency content excites simultaneously the different modes of imperfect uncertain systems and has a positive effect on their working conditions.

2.5   Mathematical Model of Large-scale Imperfect Uncertain Systems

Imperfect uncertain systems usually are large-scale systems, i.e. systems composed by a large number of interacting imperfect parts [Braiman *et al.* (1995)]. Clearly, the system size has a fundamental impact on the mathematical model, as the number of state variables increases and how the parts interact with each other needs to be defined. Let us consider the model of a one-dimensional array of coupled imperfect uncertain systems. It can be written as follows, where, to point out the peculiarity and importance of imperfect dynamics, coupling is considered to act diffusively on the dominant state variable of the imperfect dynamics:

$$
\begin{aligned}
\dot{x}_i &= f_i(x_i, \tilde{x}_i, u, w_i, p_i, t) + D_i(\tilde{x}_{i-1,j} - \tilde{x}_{i,j}) + D_i(\tilde{x}_{i+1,j} - \tilde{x}_{i,j}) \\
\dot{\tilde{x}}_i &= \tilde{f}_i(x_i, \tilde{x}_i, u, w_i, p_i, t) + \tilde{D}_i(\tilde{x}_{i-1,j} - \tilde{x}_{i,j}) + \tilde{D}_i(\tilde{x}_{i+1,j} - \tilde{x}_{i,j})
\end{aligned}
\tag{2.7}
$$

where i is the spatial index indicating the i-th imperfect uncertain system, and j indicates the j-th components of imperfect dynamics state.

This modeling approach provides a flexible mathematical representation of large-scale imperfect uncertain systems and can be easily extended to n-dimensional structures of coupled systems. The model reported in Eqs. (2.7) also fits the case of spatially-extended systems with open-loop classical vibrational control.

For our application, it is more convenient from a practical point of view to rewrite Eqs. (2.7) as:

$$
\begin{aligned}
\dot{x}_i &= f_i(x_i, \tilde{x}_i, w_i, p_i, t) + D_i(\tilde{x}_{i-1,j} - \tilde{x}_{i,j}) + D_i(\tilde{x}_{i+1,j} - \tilde{x}_{i,j}) \\
\dot{\tilde{x}}_i &= \tilde{f}_i(x_i, \tilde{x}_i, u, w_i, p_i, t) + \tilde{D}_i(\tilde{x}_{i-1,j} - \tilde{x}_{i,j}) + \tilde{D}_i(\tilde{x}_{i+1,j} - \tilde{x}_{i,j})
\end{aligned}
\tag{2.8}
$$

where the control signal $u(t)$ is injected only in the imperfect dynamics of each subsystem, so that only particular spatial points of the structure are actually excited, favoring the emergence of the hidden dynamics. Therefore, the control law is effective once the following items are specified:

- the choice of a suitable position for the actuators;
- the choice of a suitable position for the sensors;
- the choice of a suitable level of the control power;
- the appropriate on-off switching law of the action.

Chapter 3

Overview of the Electromechanical Structures

3.1 Model of the Coil

The considered systems are essentially based on a low weight mechanical structure [Hurmuzlu and Nwokah (2001)] that supports very simple rotating coils. With this term we indicate coils realized with few turns of a copper wire, having a diameter of 20mm, as the one reported in Fig. 3.1. Neodymium magnets (N45) are associated to the coils and are located below each slot, at a distance of about 1cm, providing a magnetic field of intensity $B \approx 0.03$T [Moskowitz (1995)]. A schematic representation of the single rotating coil, along with the geometric parameter values and the associated magnet, is reported in Fig. 3.2.

Fig. 3.1 Realization of a coil.

3.1.1 *Constructive details of the single coil-magnet system*

The coils must be considered nonlinear uncertain systems, due to the practical difficulties in realizing identical coils. Each coil has been realized individually, by wrapping 10 turns of 0.5mm diameter copper wire, covered by a thin layer of phenolic resin, on a temporary circular support of 18mm in

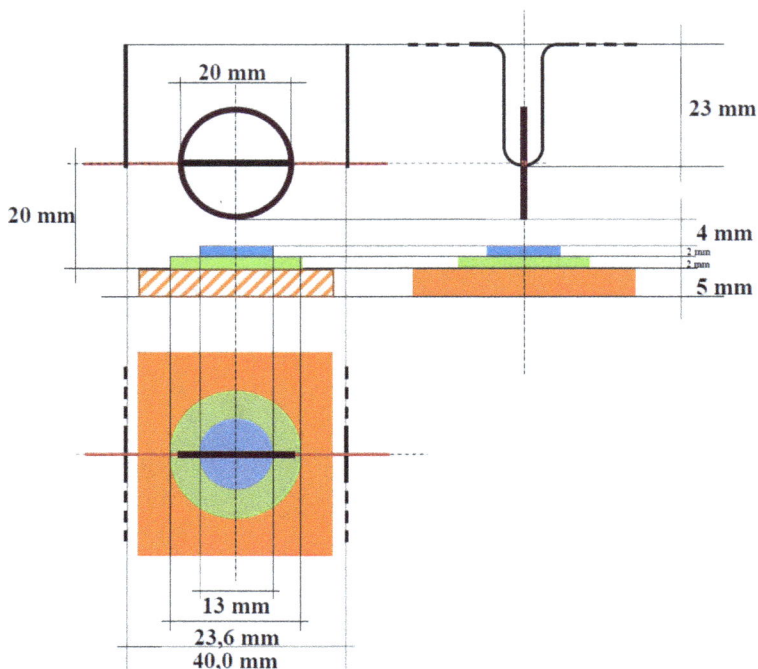

Fig. 3.2   Schematic representation of the single coil system, neodymium magnet is represented in blue, the magnet support in green and the plastic beam in orange.

diameter, so as to obtain a winding with an overall diameter of 20mm, for a total length of the conductor equal to 65cm. The rotation axis of the coil and its electrical supply system are realized by means of two copper wires of length of 2cm, and having a section with a diameter of 1mm, placed at the two extremes of the electrical turning.

The support-power system, made by the two wires connected to the coil, has been conceived in such a way as to maximize the interaction of the coil with the magnetic field generated by the permanent magnet placed immediately under it. The rigid conductor, which rests on the negative pole of the coil, is completely free from the initial insulation phenolic resin, exposing the conductive part along the entire lateral surface.

On the second support, the insulating coating was preserved for a semi-longitudinal side surface, and the rest was completely removed, exposing the copper conductive part. The two semi-longitudinal areas of the support are arranged in such a way as to modify the interaction of the field generated from the coil with the one generated by the permanent magnet alternatively at null or maximum value.

When the coil is oriented in the direction orthogonal to the magnet, it can be either or not in conduction. In conduction, the current, flowing through the coil, generates a field whose maximum flux is a vector which, in the best conditions, turns out to be exactly perpendicular to that produced by the permanent magnet. This configuration produces a torque able to induce to the coil motion around the support.

During the rotation, the support, after the semi-conductive area, meets the semi-insulated area, and, at this point, the coil is off. The interruption of the current determines the absence of the field generated by the coil and the end of the interaction with the magnetic field produced by the permanent magnet. In this operating region, the coil, winning the resistance due to friction, continues its rotatory motion due to inertia and to the angular speed imparted by the previous acceleration phase. As long as the support of the coil is powered, the alternating on/off cycle repeats producing a consequent continuous rotation of the coil.

The permanent magnet, positioned below the coil, is a sintered disc of Neodymium-Iron-Boron (NdFeB), coated with a layer of Nickel (Ni-Cu-Ni). The disk has a diameter of 13mm, the height of 2mm, the weight of 2g, and the axial direction of the magnetization (parallel to height). The quality of the magnetic material is N45 (N stands for maximum operating temperature, i.e. 80°C; 45 stands for maximum energy product, that is, the maximum amount of magnetic energy stored, i.e. 45MGOe, equal to 358kJ/m^3), with a residual magnetism $B_r = 1.35$T, and a coercive field strength of flux of density $bH_c = 995$KA/m and of polarization $jH_c = 955$KA/m. These values, along with the weak field produced by the rotating coil above the permanent magnet, ensure that the magnet itself is quite insensitive to the variation of the characteristics due to the interaction with external fields. Therefore, the magnetic performance can be considered constant throughout the operation.

The magnet rests on a soft galvanized iron disc of a diameter of 23.6mm and a thickness of 2mm, which essentially has the purpose of making the support of the magnet stable.

To give an idea of the intensity of the magnetic field generated by the permanent magnet, to which the coil is subject, we can say that a magnet with the same characteristics, placed at a distance of 16mm, equal to the axis distance of the coil from the top face of the magnetic disk, is attracted with a force of about 0.235N.

3.1.2   *Mathematical model of the single rotating coil*

The i-th coil is described by the following nonlinear dynamical model [Buscarino *et al.* (2016b,c)]:

$$\begin{cases} \dot{X} = Y \\ \dot{Y} = \frac{1}{J}(ISB\sin(X) - KY) \end{cases} \tag{3.1}$$

where X represents the phase of the coil, Y the angular speed, J the angular momentum, and K the damping factor. I is the current flowing into the coil and is given by

$$I = \frac{V_a - YSB\sin(X)}{R(X)}$$

with V_a the voltage supplied to the coil, S the coil area, B the magnetic field and $R(X)$ the contact resistance, which, due to the coil construction constraint, is nonlinear according to:

$$R(X) = \begin{cases} 0.2\Omega \ \text{if } X < \pi \\ 10k\Omega \ \text{if } X \geq \pi \end{cases}$$

As concerns the friction factor K between the coil and its support, we adopted a nonlinear switching friction model [Karnopp (1985)] on the basis of which friction coefficient depends on the angular speed of coil. In particular, we have selected

$$K = \begin{cases} K_H \ \text{if } Y < 2\text{Hz} \\ K_L \ \text{if } Y \geq 2\text{Hz} \end{cases}$$

The system in Eqs. (3.1) can be numerically integrated fixing parameter values according to nominal values estimated for a single coil. We fixed $V_a = 2\text{V}$, $B = 0.008\frac{g}{Am}$, $K_H = 10^{-3}\frac{gm^2}{s}$, $K_L = 5 \cdot 10^{-6}\frac{gm^2}{s}$, $S = 3.14 \cdot 10^{-4}m^2$, $J = 6 \cdot 10^{-6}gm^2$ and obtained the trend of the angular speed reported in Fig. 3.3(a). Furthermore, the current I is reported in Fig. 3.3(b). It can be noticed that, with this set of parameter values, the coil rotates with a mean angular speed of 20Hz, which is in the range of the typical coils used in the experiment.

(a)

(b)

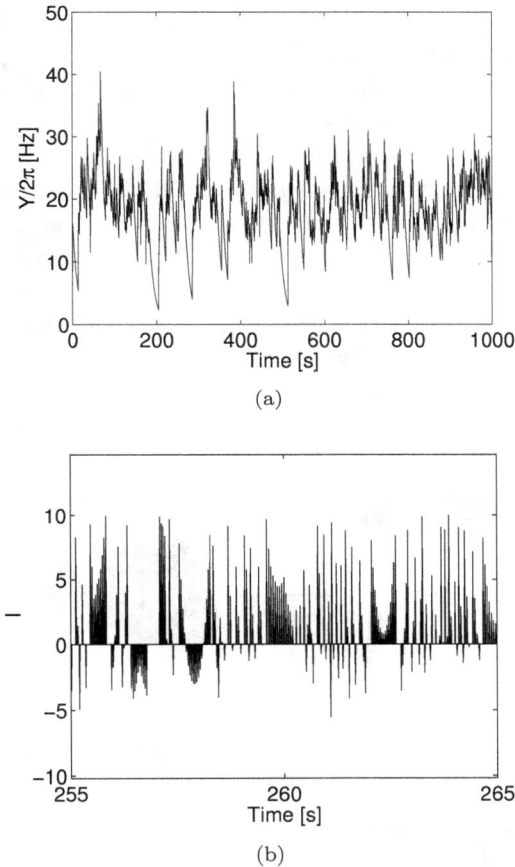

Fig. 3.3   Numerical integration of the model in Eqs. (3.1): (a) angular speed of the coil in hertz, (b) current flowing in the coil.

3.1.3   *Sensors equipments for coil model validation*

The system that allows to measure its angular speed is fundamental to validate the model formulated for the coil. The tachometric system must be reliable, providing a measurement as accurate as possible, and, at the same time, not affecting the rotatory motion of the coil and its dynamics. In the choice of the measuring system, mechanical or magnetic sensors have been excluded, since their adoption would introduce a significant interaction with the dynamical behavior of the coils. The choice settles on a reflection optical system capable of retrieving the passage of a reference element located on the coil itself. In such a system, in order to avoid that the ambient light influences the measurement, we have chosen a light source with a wavelength far from the visible range. In choosing the suitable sensor, the cost of the equipment has been also taken into account.

In order to fulfill all the requirements, we selected a reflective optical sensor with transistor output, namely the TCRT5000 produced by Vishay Semiconductors [Webster and Eren (2014)], shown in Fig. 3.4. This device is very compact in size ($L \times W \times H$ in mm: $10.2 \times 5.8 \times 7$), robust in operation and, in the same package, contains the infrared photodiode emitter and the detector phototransistor. The latter is provided with filter block for the ambient light in the visible. The functioning principle of the system is very simple.

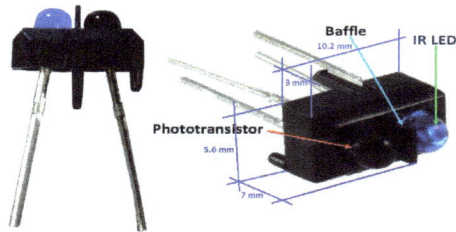

Fig. 3.4   Schematic representation of the TCRT5000 optical sensor.

The photodiode, suitably powered, emits in the near infrared (NIR or IR-A), with a wavelength of 950nm. The phototransistor, housed into the container space, is separated from the source by a baffle which prevents the direct infrared radiation to reach and be activated continuously. Inside the TCRT5000 in both active parts (emitter and detector), lenses are incorporated. The angular characteristics of both are divergent. In the TCRT5000, the concentration of the beam reaches an angle of 16° for the emitter and 30° for the detector, respectively, resulting in operation on an increased range with optimized resolution.

If a reflective surface is located on the rotating coil, the light is emitted by the diode, reflected, and then detected by the phototransistor. When the phototransistor is not illuminated it is in interdiction, on the contrary, if it is reached by the infrared radiation, a current flows from the collector to the base by turning on the phototransistor in a stable and continuous manner, up to the persistence of the infrared excitation.

The parameter that describes the operation of the optical coupling precisely is the so called optical transfer function (OT) of the sensor. It is defined as the ratio of the received to the emitted radiant power

$$OT = \frac{\Phi_r}{\Phi_e}$$

The operating range, the resolution of the object optical distance, the sensitivity and the switching point in case of local variations in reflection, which constitute further parameters of the sensor, are directly related to this optical transfer function.

In the case of reflective sensors with phototransistors as receivers, such as the TCRT5000, the ratio I_C/I_F, i.e. the ratio between the collector current $I_{out} = I_C$ and the forward current $I_{in} = I_F$ of the emitting diode is preferred to the optical transfer function. This ratio is generally known as the coupling factor, $k = \frac{I_C}{I_F}$. The following approximate relationship exists between k and OT:

$$k = \frac{SB}{h} \frac{\Phi_r}{\Phi_e}$$

where B is the current amplification, $S = \frac{I_h}{\Phi_r}$ is the phototransistor spectral sensitivity, and $h = \frac{I_F}{\Phi_e}$ is the proportionality factor of the transmitter.

Beyond the forward current, I_F, and the temperature, the coupling factor also depends on the distance from the reflecting surface and the frequency, that is, the speed of reflection change. Very often, in the data sheets, rather than k, reference is made to the current transfer ratio (CTR), which is expressed as a percentage indicating the relationship between the output current and the input current. The CTR is a parameter similar to the DC current amplification ratio of a transistor (h_{FE}).

The response time of a reflective sensor is proportional to the load resistance, so that for high-speed signals the load resistance must be designed as small as possible within the allowed range. However, when the load resistance is minimized, the transistor may not turn completely on and the output signal may be unstable. Since the response time of the optical system is not a critical element for our tachometric system, the frequencies of rotation of the coils being in the order of a few tens of hertz, the correct operation of the sensor was achieved by optimizing the CTR, according to the manufacturer guidelines.

Furthermore, the current on the LED has to be kept minimal so that spurious signals, which would provide phototransistor erroneous activations compromising the measure, are limited. Given the plane of the coil, the sensor has been positioned with its axis emitter/detector orthogonal to the coil itself, in such a way that the separation baffle between the emitter and the detector lies on the plane perpendicular to that of rotation. This configuration allows the upper part of the loop to reflect the maximum amount of infrared radiation, when it is exactly positioned transversely to the sensor, and, with its rotation, to generate a signal that varies without abrupt changes. The current flowing through the photo-transistor follows the dynamics of the reflection of infrared radiation, reaching its maximum value in correspondence with the coil passing in proximity to the sensor separation baffle. The location of the sensor with respect to the coil is shown in Fig. 3.5.

Fig. 3.5   Location of the optical sensor above the rotating coil.

From the data provided by the manufacturer, it emerges that, for an operating distance in the range $[0.2 \div 15]$mm, its collector current is greater than 20%, while the optimum working distance is indicated in 6.5mm, therefore, even if these values are calculated under ideal test conditions, they allow us to locate the sensor at a distance from the coil such as to accurately infer the speed data and to derive the dynamics of the rotating system without significant interference.

The electronic circuit realized for conditioning the sensor is reported in Fig. 3.6.

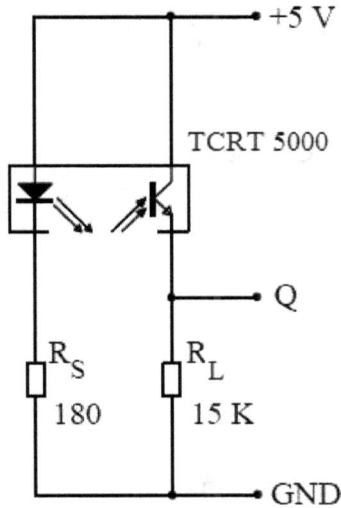

Fig. 3.6   Conditioning circuit for the optical sensor.

The configuration chosen for the phototransistor with emitter resistance, ensures that the output voltage is very small, close to zero, for all the time that the upper part of the coil is positioned far from the IR source. In the proximity of the LED, the coil reflects an ever-increasing amount of IR radiation and this provides an increase of the current between the collector and the base in the phototransistor, contributing to the emitter current flow, and, therefore, increasing the voltage across the resistor R_L. In this way a variable signal whose peaks follow the transits of the upper edges of the coil in the proximity of the sensor is obtained. The time interval between two successive peaks is equal to the half-period of coil rotation around its axis, as shown in the schematic representation of Fig. 3.7.

The sensor output, therefore, is a voltage which can be acquired through a data-acquisition (DAQ) board in a computer, or simply observed by means of an oscilloscope. A typical situation is reported in Fig. 3.8. It refers to the output V_o of an optical sensor monitoring the rotation of a single coil. Each spike corresponds to the passage of the coil nearby the sensors, therefore for each two spikes the coil performs a complete rotation. The angular speed of the coil can be derived calculating the distance between a peak and the second one after it. According to data tips displayed in Fig. 3.8, the angular speed in the given time window is $\frac{1}{15.105-15.063} \approx 23.81 \text{Hz}$.

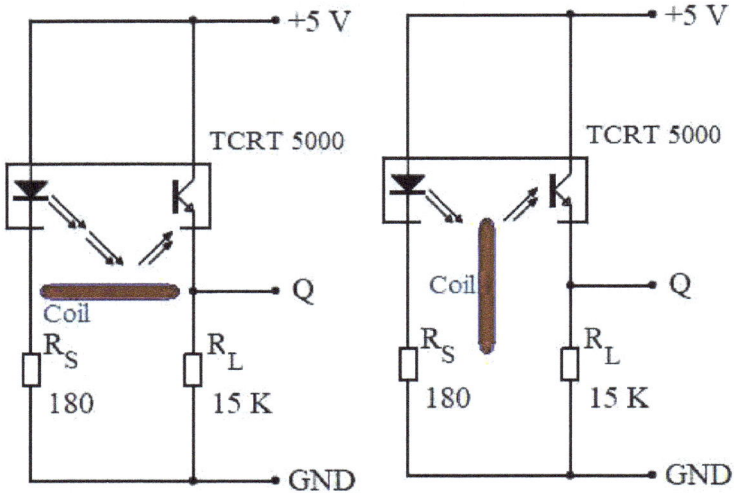

Fig. 3.7   Schematic representation of the sensor working principle: (a) the coil reflects the IR outside the phototransistor visibility field, (b) the coil reflects the IR on the phototransistor.

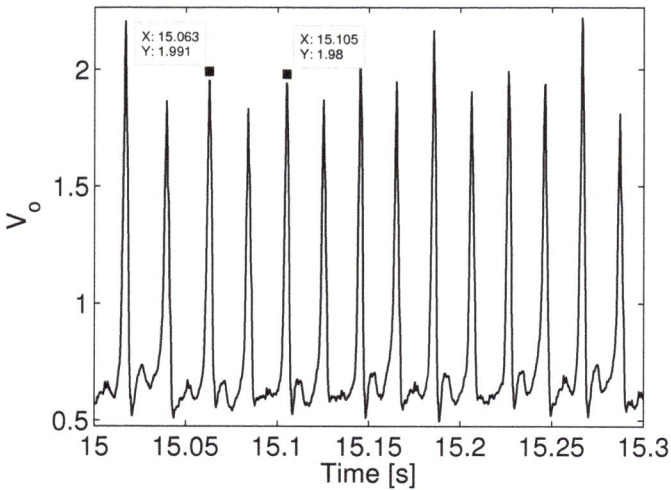

Fig. 3.8   Output of the optical sensors during the rotation of the corresponding coil.

The values assigned to the conditioning circuit components are derived essentially according to the desired values of the current on the LED and the one that must flow through the collector. The data are taken from charts supplied by the manufacturer, in accordance with the need to optimize the operation and the sensor performance, in terms of optical coupling, operating temperature, stability and switching speed.

For a minimum distance of the coil from sensor of about 6.5mm, corresponding to the point of maximum reflection provided by the coil in the infrared range, according to sensor data sheet it is assumed a current $I_F = 20$mA, which, in relation to the sensor specifications, assures, in ideal tests, the best conditions of coupling, the parameter k value has its maximum and the LED self-heating is minimal because of the low power dissipated (about 50mW up to 40mA).

The power emitted by the LED is about 55μW, this allows us to conclude that the action of the ambient light can be neglected since it can be quantified in 0.025μW. Given the variability of optical coupling of the system, it is not easy to estimate *a priori* the current I_c in the collector, this must be experimentally detected.

The switching frequency is also well within the operating specifications, as with a $+5$V supply voltage and a $R_L = 15$kΩ, we can estimate a frequency of 1.2kHz which is three orders of magnitude higher than the rotation frequency of the coil.

3.2 Mechanical Supports

3.2.1 *General description*

The structures considered are of the type shown in Fig. 1.8. They consist of two trails (rectangular or circular) with a set of N slots which host the coils. The trails also allow electrical power to be supplied to the coils. Magnets associated to each coil are located on the bottom part of the setup, as shown in Fig. 3.9.

These structures belong to the class of large-scale dynamical systems, in the sense that the low weight of the structure, including the coils, is negligible with respect to the volume. The first of the tasks that we want to achieve is the complete start-up of the various coils, i.e. the onset of an effective and steady rotation. As a second task, the regularization (synchronization) of the angular speed of the coils is accounted for. As we will discuss in the following, in fact, both tasks are nontrivial.

Fig. 3.9   Detail of the coupling permanent magnet-coil.

The mechanical supports reported in Fig. 1.8, which from now on are referred to as rectangular structure (Fig. 1.8(a)) and circular structure (Fig. 1.8(b)), are flexible structures with several natural modes. The coils, ideally represented by the model in Eqs. (3.1), as mentioned, are nonlinear uncertain systems, due to the practical difficulties in realizing identical coils.

Another source of uncertainty derives from the fact that the coils are located in the mechanical structure, which is free to move in all three directions. Moreover, the slots where the coils are located, shown in Fig. 3.10, are different from each other. All these factors contribute to create adverse conditions to attain a regular behavior of the system. In particular, depending on the initial conditions and system parameters, when the power is switched on, typically it occurs that some coils start rotating, while others do not.

The idea is to design flexible structures that, mechanically excited by their self-oscillations, allow the coils which are not working to receive

Fig. 3.10   Detail of the slot, where the coils are located, in rectangular structure.

solicitations making them able to overcome the initial (electromechanical) inertia and start rotating. This should occur through a global action derived from the mechanical vibrations that couple coils with each other. Therefore, the mechanical vibrations should favor the self-organization of the system in order to overcome the effects of the uncertainties.

As far as the start-up phase is concerned, the two structures of Fig. 1.8 show a similar behavior. In particular, coils in both structures are mechanically and electrically coupled and, therefore, from the mechanical point of view the elasticity favors vibrations. If the electrical power is applied, the start-up phase is critical and the self-generated vibrations in the system are not enough to let the complete start-up of all the coils. In order to overcome this situation, a shock system actuated with a low power electromechanical actuators has to be used with the aim of exciting the electromechanical structures to favor active vibrations and, therefore, to move the coils towards the condition to start-up.

3.2.2 *Construction details of the rectangular and circular structures*

We focus our attention now on the realization of the two electromechanical structures. Both of them are composed by a copper string of 1mm width covered by a layer of resin which plays the double role of providing electrical isolation and preserves from the oxidation of the structure. The mechanical characteristics and the metallurgic properties of copper allow to easily bend the strings and shape the structures maintaining a high level of robustness, elasticity and electrical conductibility [Feodosev (1977)].

The rectangular structure, characterized by the physical parameters schematized in Fig. 3.11, is constituted by two parallel portals (28cm long) whose extreme points are linked by two ties, each one with a length of 5cm. The two portals are supported and anchored to the base by four piers (17cm height) located at the extreme points. A further pair of piers is located in the middle of the portals, and, as shown in Fig 3.12, are equipped with two springs. These two piers have the crucial role of avoiding the bending of the portals due to the total weight of the coils (18g).

A series of ten supports, at a distance of 2.5cm from one other, are realized over the portals bending the copper strings to a U shape where each side is 1.5cm long and the spacing between them is 8mm. Over the curvature of the bending, the isolating resin has been stripped to ensure the electrical conductibility providing power to the coil. As it is shown in

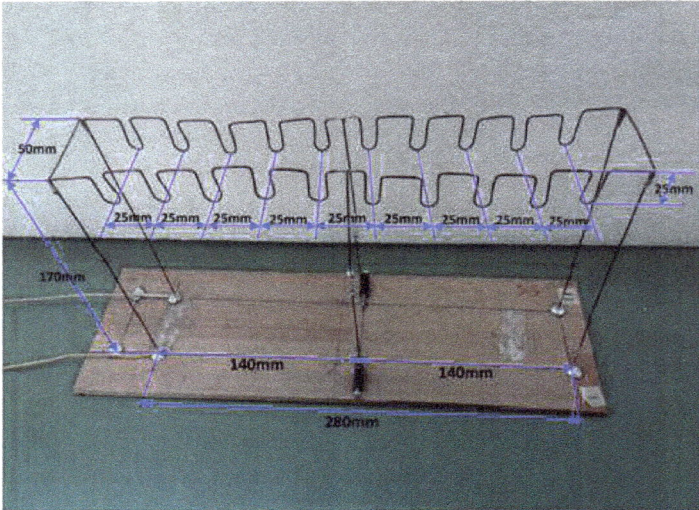

Fig. 3.11   The rectangular structure used for the experiments. Physical dimensions are indicated.

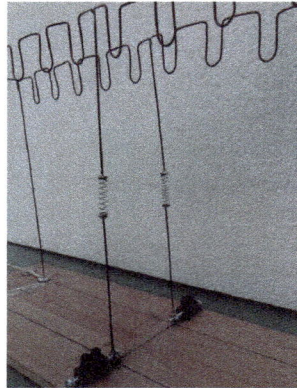

Fig. 3.12   Detail of the springs located in the central imposts in order to sustain the portals and without affecting the elasticity of the structure.

Fig. 3.11, the geometry of the supports is clearly subjected to uncertainty and imperfections on which the actual working conditions of the whole structure deeply rely, as it will be discussed later on.

As concerns the circular structure, whose geometric parameter values are reported in Fig. 3.13, it is constituted by three concentric rings equipped with vertical piers. Coils are allocated between the rings with a central symmetry. Each vertical pier has length of 17cm, and it is anchored to the base. The use of piers is fundamental ensuring the possibility to have oscillations over planes orthogonal to the coils rotation and allows for torsional oscillations around the central axis. The three rings are spaced by $R = 4.5$cm. The radius of the internal ring is $R_1 = 4.5$cm, the middle ring has a radius $R_2 = R_1 + R = 9$cm, the external ring has a radius $R_3 = R_1 + 2R = 13.5$cm.

A total number of 16 supports has been realized through the two circular sectors marked by the three rings, spaced by an angular distance of 22.5°. The electrical power is supplied to the structure by means of external power suppliers providing a fixed voltage $V_a = 2V$ and a maximum current $I_a = 3A$. The negative terminal is provided to the central ring, while the inner and the outer ones are connected to the positive terminal.

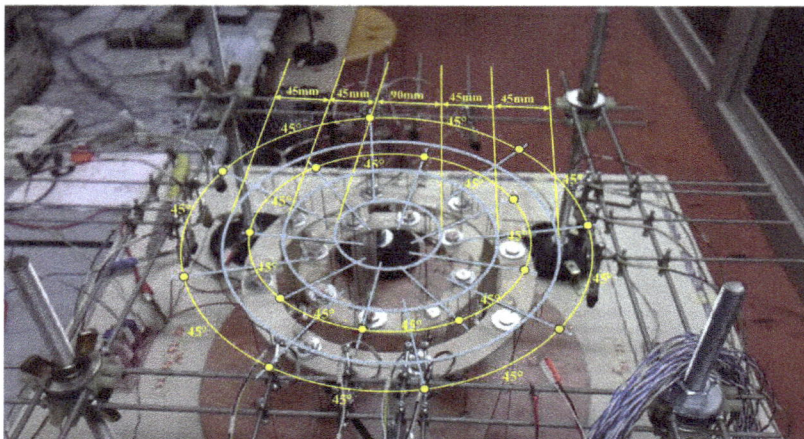

(a)

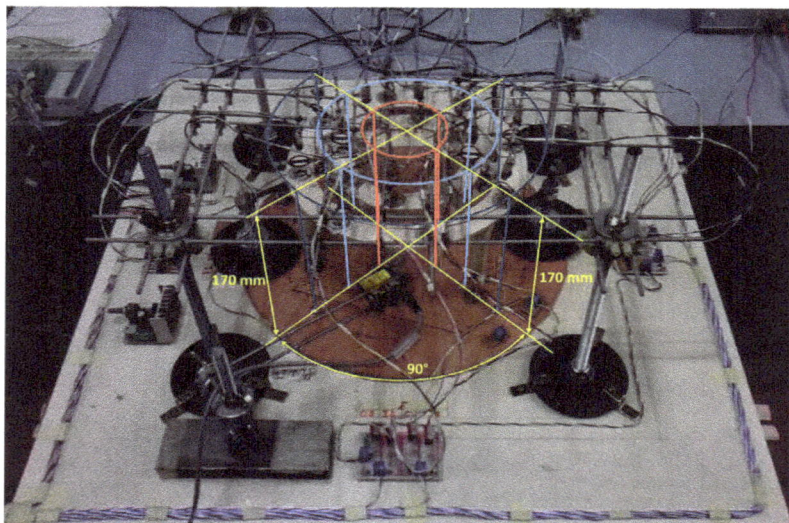

(b)

Fig. 3.13   The circular structure used for the experiments.   Physical parameters are indicated: (a) top of the structure, (b) side of the structure.

Chapter 4

Large-scale Electromechanical Systems with Passive and Active Vibrations

As already mentioned, technological imperfections on the coils and on the supporting structures lead to the fact that coils are unable to overcome the initial electromechanical inertia and so start rotating. On the other hand, the system may be considered to properly operate if and only if, after a transient time, all the coils are rotating. We remark that, if the power supply is sufficiently high, many coils will be operating because many of them will have a self-sustained start-up and will induce in the mechanical structure a passive vibration that may change the angular position of the other coils and let them start.

In this case the system will properly operate thanks to passive vibrations. From an experimental point of view by using unlimited energy, the event that in steady state some coils are not working is a rare one. In the case that the energy is limited (the case of power energy supply optimization) it is common to find many coils not operating and, on the contrary, it is a rare case that the system is self starting. This is the most interesting and realistic (as energy is limited) case, which requires the use of an external actuation, like that shown in Chapter 2, in accordance with the control scheme of Fig. 2.4. As introduced in Chapter 2, the control scheme operates in closed loop. For our specific case study, the control feedback makes use of information on the angular speed of the coils. In particular, the angular speed of the two coils at the extreme ends of the structure are the more critical, due to the mechanical stiffness of the part where they are located. Therefore, the use of the signal related to the angular speeds of these two coils is more convenient to close the control loop. The control will be performed until the two highly sensitive coils begin to rotate. Being the most difficult coils to be controlled, their proper operation assures that all systems are working.

In the numerical simulations discussed in this chapter, we are interested in verifying the possibility to excite the imperfect dynamics, therefore the feedback is not implemented, but considered always active. Furthermore, we stress the fact that despite explicitly accounting for parameters uncertainty, the beneficial role in terms of both start-up and regularization is only due to the excitation of the imperfect dynamics.

4.1   Modeling the Rectangular Structure

The mathematical model of the system of Fig. 1.8(a) (with $N = 10$ coils) with the actuated control signal written in accordance with the general framework presented in Chapter 3 is now introduced:

$$\begin{cases} \dot{X}_i = Y_i \\ \dot{Y}_i = \frac{1}{J_i}(I_i S_i B_i \sin(X_i) - K_i Y_i) + \bar{D}_i(\tilde{x}_{i+1,1} - \tilde{x}_{i,1}) + \bar{D}_i(\tilde{x}_{i-1,1} - \tilde{x}_{i,1}) \end{cases}$$
$$(4.1)$$

with

$$\begin{cases} \dot{\tilde{x}}_{i,1} = \alpha_i \tilde{x}_{i,2} + u(t) \\ \dot{\tilde{x}}_{i,2} = -\alpha_i \tilde{x}_{i,1} \end{cases} \qquad (4.2)$$

where α_i is the natural frequency of the imperfect dynamics associated to the i-th coil ($i = 1, \ldots, 10$), $\bar{D}_i$ are the diffusion coefficients associated to the imperfect dynamics. The signal $u(t)$ is the combined effect of the control action on the system. We emphasize that the angular momentum, coil area, magnetic field and contact resistance are now indexed with a subscript to account for uncertainty in the system.

The friction factor K_i between each coil and its support depends on the angular speed of each coil as

$$K_i = \begin{cases} K_{H,i} & \text{if } Y_i < 2\text{Hz} \\ K_{L,i} & \text{if } Y_i \geq 2\text{Hz} \end{cases}$$

To take into account the role of uncertainties in the model, we fixed parameters J_i, S_i, B_i, $K_{H,i}$, and $K_{L,i}$ as random variables taken from a Gaussian distribution with given mean value and variance σ^2.

4.1.1   *Simulation of the start-up phase*

In Fig. 4.1, the results of a statistical analysis on the start-up behavior of the model in the presence of uncertainties are reported. We considered different variances for the Gaussian distribution of the friction coefficient $K_{L,i}$, while other parameters have been set as reported in the figure caption.

In Fig. 4.1(a) the probability of having N_R rotating coils is shown over 500 trials when the term $u(t)$ in Eqs. (4.2) representing the control input to elicit the imperfect dynamics is set to zero, i.e. no control action is applied. As it can be observed, the probability of observing $N_R > 4$ rotating coils is close to zero, independently from the variance of the friction factor. When a control signal $u(t)$ is provided to the imperfect dynamics, the probability of observing all the coils actually rotating is sensibly increased, as shown in Fig. 4.1(b).

4.1.2 *Simulation of the control phase (regularization)*

Let us suppose that the start-up phase has been correctly completed, therefore all coils are actually rotating and the structure vibrates mainly as a consequence of the imperfect dynamics. A low level control signal is still active, i.e $u(t) = 0.1 \sin(10t)$ with the aim of favoring the self-organizing behavior exploiting the role of imperfections.

In Fig. 4.2(a) the temporal evolutions of the angular speeds Y_i of the ten coils are reported for the case of identical systems. In this case, we fixed $V_a = 2V$, $B = 0.03T$, $\bar{D} = 10$, $K_{H,i} = 10^{-3}$, $K_{L,i} = 2 \cdot 10^{-5}$, $S_i = 5 \cdot 10^{-5}$, $J_i = 6 \cdot 10^{-6}$, and $\alpha_i = 10 \text{rad/s}$ for all $i = 1 \ldots 10$. The angular speeds display some mismatches, especially among those located in the central region of the system. This means that the control action alone is not able to provide the desired effect.

We now consider the inclusion of a degree of uncertainty, namely drawing at random J_i, S_i, $K_{L,i}$ and α_i from a Gaussian distribution with $\langle J_i \rangle = 6 \cdot 10^{-6}$, $\sigma(J_i) = 10^{-6}$, $\langle S_i \rangle = 5 \cdot 10^{-5}$, $\sigma(S_i) = 10^{-6}$, $\langle \alpha_i \rangle = 10 \text{rad/s}$, $\sigma(\alpha_i) = 0.5$, $\langle K_{L,i} \rangle = 2 \cdot 10^{-5}$, $\sigma(K_{L,i}) = 10^{-6}$. From visual inspection of Fig 4.2(b), reporting the angular speeds Y_i of the ten coils, it is evident that the interplay between the low level control signal exciting imperfections and the presence of uncertainties leads to a regularization.

4.2 Modeling the Circular Structure

The main difference between the rectangular and the circular structure is the fact that each coil is located within two other coils. The rectangular structure, in fact, is a one-dimensional open array, while the circular structure is a ring, i.e. a closed chain. Therefore, to simulate the behavior of the circular structure by means of model (4.1) it is sufficient to change the boundary conditions, allowing the diffusion between the last and the first cells.

4.2.1   *Simulation of the start-up phase*

We performed the same statistical analysis for the start-up behavior of the model in the presence of uncertainties discussed above for the rectangular structure and report the results in Fig. 4.3. We considered again different variances for the Gaussian distribution of the friction coefficient $K_{L,i}$, while the other parameters have been set as reported in the figure caption. In Fig. 4.3(a) the probability of having N_R rotating coils is shown over 500 trials when the term $u(t)$ in Eqs. (4.2) is set to zero. As it can be observed, the probability of having $N_R > 3$ rotating coils is negligible, independently from the variance of the friction factor. When the control signal $u(t)$ is provided to the imperfect dynamics, the probability of observing all the coils actually rotating is almost 100%, as shown in Fig. 4.3(b).

4.2.2   *Simulation of the control phase (regularization)*

We now focus on the regularization of angular speeds, considering that all coils are rotating over their support. A low level control signal is activated, i.e. $u(t) = 0.1\sin(15t)$ in order to excite imperfections and their role on self-organization.

Let us start from the case of identical systems, discussed in Fig. 4.4(a). We fixed $V_a = 2V$, $B = 0.03T$, $\bar{D} = 10$, $K_{H,i} = 10^{-3}$, $K_{L,i} = 2 \cdot 10^{-5}$, $S_i = 5 \cdot 10^{-5}$, $J_i = 6 \cdot 10^{-6}$, and $\alpha_i = 15\text{rad/s}$ for all $i = 1 \ldots 10$ and simulate the system retrieving inhomogeneous angular speeds.

Introducing a source of uncertainty, the low level control signal is able to excite the imperfect dynamics and perform the desired regularization. In fact, considering J_i, S_i, $K_{L,i}$ and α_i drawn at random from a Gaussian distribution with $\langle J_i \rangle = 6 \cdot 10^{-6}$, $\sigma(J_i) = 10^{-6}$, $\langle S_i \rangle = 5 \cdot 10^{-5}$, $\sigma(S_i) = 10^{-6}$, $\langle \alpha_i \rangle = 10\text{rad/s}$, $\sigma(\alpha_i) = 0.5$, $\langle K_{L,i} \rangle = 2 \cdot 10^{-5}$, $\sigma(K_{L,i}) = 10^{-6}$, the angular speed is more homogenous as it can be observed in Fig 4.4(b), where the state variables Y_i of the 16 coils are reported.

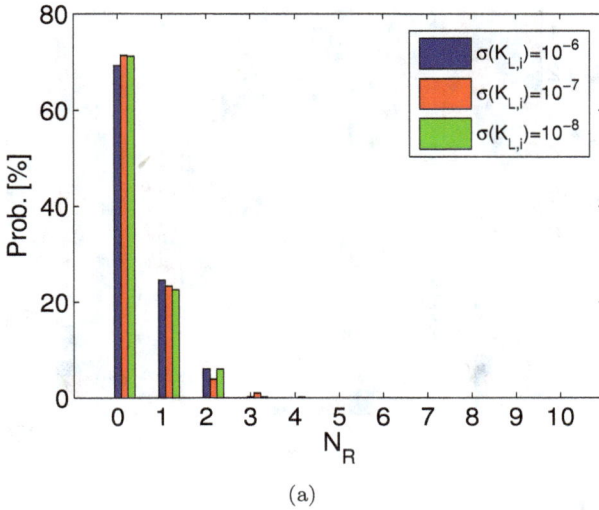

(a)

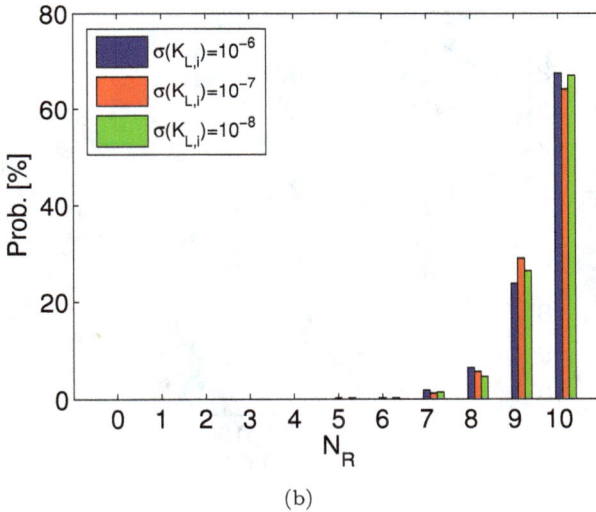

(b)

Fig. 4.1 Numerical simulation of the start-up phase in the model (4.1) for the rectangular structure: probability of observing N_R rotating coils over 500 realizations. Parameters are randomly chosen for each realization according to Gaussian distributions with $\langle J_i \rangle = 6 \cdot 10^{-6} gm^2$, $\sigma(J_i) = 10^{-6}$, $\langle S_i \rangle = 3.14 \cdot 10^{-4} m^2$, $\sigma(S_i) = 10^{-6}$, $\langle \alpha_i \rangle = 10 \text{rad/s}$, $\sigma(\alpha_i) = 0.5$, $\langle K_{L,i} \rangle = 2 \cdot 10^{-5} \frac{gm^2}{s}$, $\sigma(K_{L,i}) = 10^{-6}$ (blue histograms), $\sigma(K_{L,i}) = 10^{-6}$ (red histograms), $\sigma(K_{L,i}) = 10^{-6}$ (green histograms), and $V_a = 2\text{V}$, $B = 0.008 \frac{g}{Am}$, $\bar{D} = 10$, $K_{H,i} = 10^{-3} \frac{gm^2}{s}$. (a) Control action OFF, i.e. $u(t) = 0$ in the simulation, (b) control action ON, i.e $u(t) = 10\sin(10t)$ in the simulation.

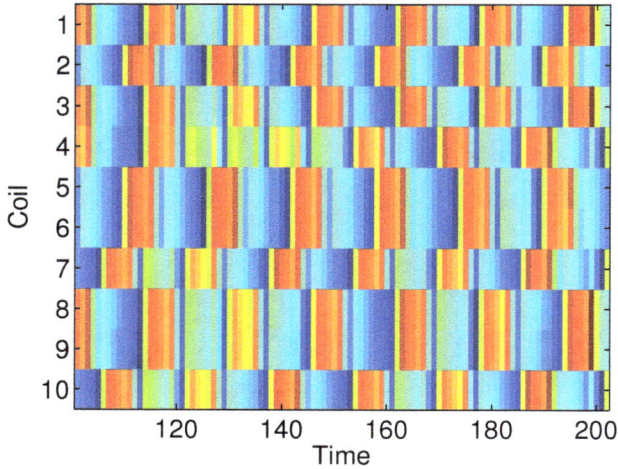

(a)

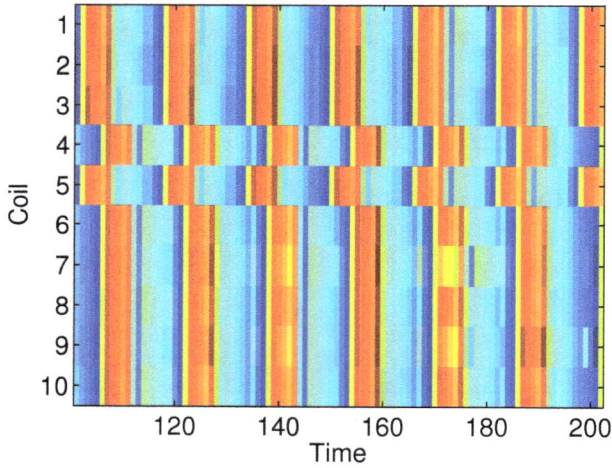

(b)

Fig. 4.2   Numerical simulation of the control phase in the model (4.1) for the rectangular structure: temporal evolution of the angular speeds for the ten coils. Parameters are chosen as: (a) $V_a = 2V$, $B = 0.008 \frac{g}{Am}$, $\bar{D} = 10$, $K_{H,i} = 10^{-3} \frac{gm^2}{s}$, $K_{L,i} = 2 \cdot 10^{-5} \frac{gm^2}{s}$, $S_i = 3.14 \cdot 10^{-4} m^2$, $J_i = 6 \cdot 10^{-6} gm^2$, and $\alpha_i = 10$rad/s for all $i = 1 \ldots 10$, (b) drawn from a Gaussian distribution with $\langle J_i \rangle = 6 \cdot 10^{-6} gm^2$, $\sigma(J_i) = 10^{-6}$, $\langle S_i \rangle = 3.14 \cdot 10^{-4} m^2$, $\sigma(S_i) = 10^{-6}$, $\langle \alpha_i \rangle = 10$rad/s, $\sigma(\alpha_i) = 0.5$, $\langle K_{L,i} \rangle = 2 \cdot 10^{-5} \frac{gm^2}{s}$, $\sigma(K_{L,i}) = 10^{-6}$. Other parameters: $V_a = 2V$, $B = 0.008 \frac{g}{Am}$, $\bar{D} = 10$, $K_{H,i} = 10^{-3} \frac{gm^2}{s}$, $u(t) = 0.1 \sin(10t)$.

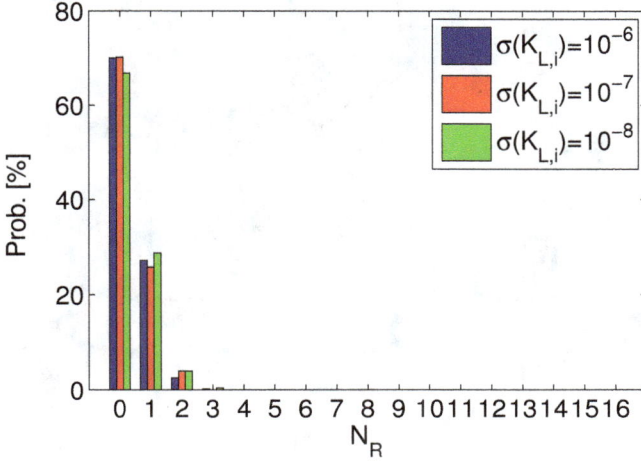

(a)

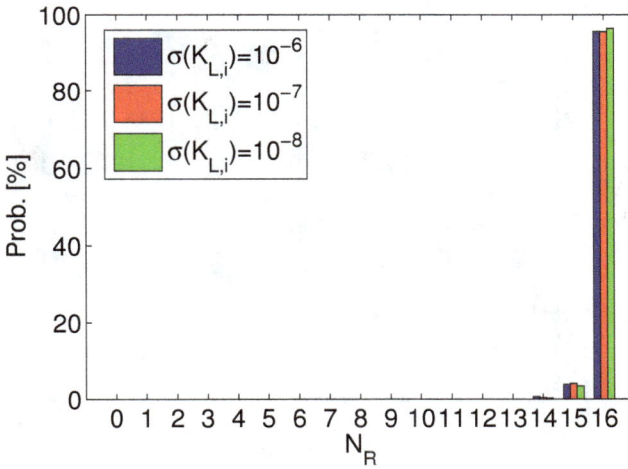

(b)

Fig. 4.3 Numerical simulation of the start-up phase in the model (4.1) for the circular structure: probability of observing N_R rotating coils over 500 realizations. Parameters are randomly chosen for each realization according to a Gaussian distribution with $\langle J_i \rangle = 6 \cdot 10^{-6} gm^2$, $\sigma(J_i) = 10^{-6}$, $\langle S_i \rangle = 3.14 \cdot 10^{-4} m^2$, $\sigma(S_i) = 10^{-6}$, $\langle \alpha_i \rangle = 10 rad/s$, $\sigma(\alpha_i) = 0.5$, $\langle K_{L,i} \rangle = 2 \cdot 10^{-5} \frac{gm^2}{s}$, $\sigma(K_{L,i}) = 10^{-6}$ (blue histograms), $\sigma(K_{L,i}) = 10^{-6}$ (red histograms), $\sigma(K_{L,i}) = 10^{-6}$ (green histograms), and $V_a = 2V$, $B = 0.008 \frac{g}{Am}$, $\bar{D} = 10$, $K_{H,i} = 10^{-3} \frac{gm^2}{s}$. (a) Control action OFF, i.e. $u(t) = 0$ in the simulation, (b) control action ON, i.e $u(t) = 50\sin(10t)$ in the simulation.

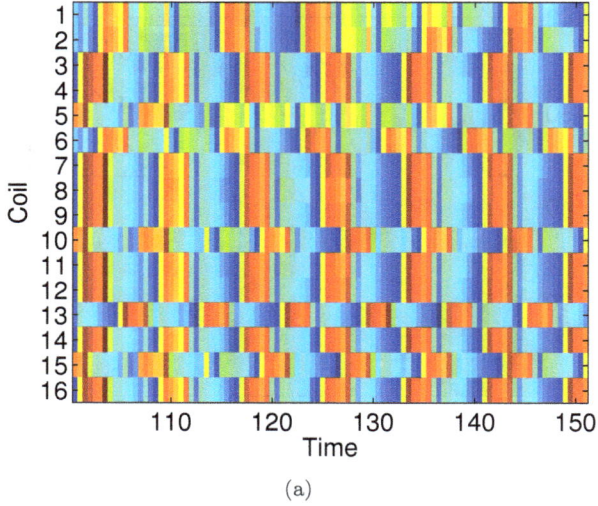

(a)

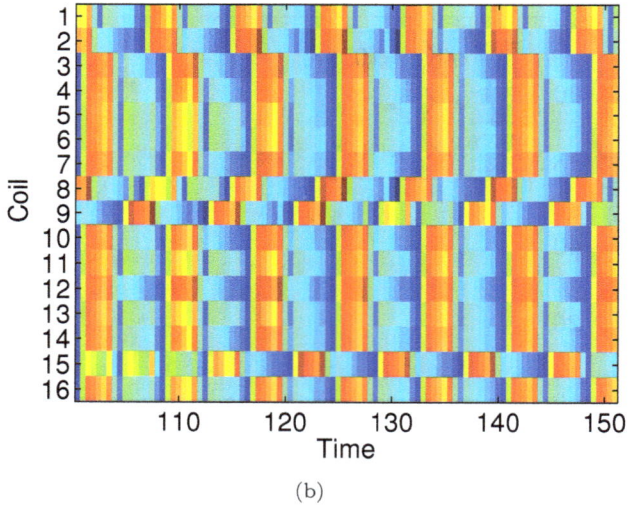

(b)

Fig. 4.4   Numerical simulation of the control phase in the model (4.1) for the circular structure: temporal evolution of the angular speeds for the sixteen coils. Parameters are chosen as: (a) $V_a = 2\text{V}$, $B = 0.008\frac{g}{Am}$, $\bar{D} = 10$, $K_{H,i} = 10^{-3}\frac{gm^2}{s}$, $K_{L,i} = 2 \cdot 10^{-5}\frac{gm^2}{s}$, $S_i = 3.14 \cdot 10^{-4}m^2$, $J_i = 6 \cdot 10^{-6}gm^2$, and $\alpha_i = 10\text{rad/s}$ for all $i = 1\ldots10$, (b) drawn from a Gaussian distribution with $\langle J_i \rangle = 6 \cdot 10^{-6}gm^2$, $\sigma(J_i) = 10^{-6}$, $\langle S_i \rangle = 3.14 \cdot 10^{-4}m^2$, $\sigma(S_i) = 10^{-6}$, $\langle \alpha_i \rangle = 10\text{rad/s}$, $\sigma(\alpha_i) = 0.5$, $\langle K_{L,i} \rangle = 2 \cdot 10^{-5}\frac{gm^2}{s}$, $\sigma(K_{L,i}) = 10^{-6}$. Other parameters: $V_a = 2\text{V}$, $B = 0.008\frac{g}{Am}$, $\bar{D} = 10$, $K_{H,i} = 10^{-3}\frac{gm^2}{s}$, $u(t) = 0.1\sin(15t)$.

Chapter 5

Equipments and a Gallery of Experiments

5.1 Overview of the Experimental Setup

A complete overview of the experimental setup designed and implemented for the study of the electromechanical systems discussed before is shown in Fig. 5.1.

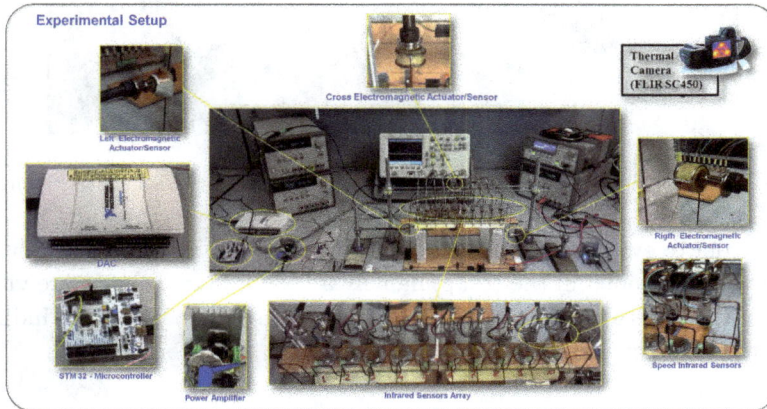

Fig. 5.1 Complete system setup for the rectangular structure including virtual instruments and infrared thermal camera to monitor coil angular speed, and control modules implemented on a micro-controller.

The elements involved in the setup include the data-acquisition board National Instrument NI-USB6255 performing the digital to analog conversion of the measurements with a resolution of 16 bits and a sample rate of 250kS/s. Moreover, the arrangement of TCRT5000 reflective optical sensors with transistor output, described in Chapter 3, is displayed for the

49

rectangular structure. The same sensors are also adopted in the circular structure, arranged according to Fig. 5.2.

Fig. 5.2   Experimental setup, including sensors, related to the circular structure.

In order to evaluate the behavior of the structures, further measurement systems are included in the setup, namely a laser system to measure vibrations of the piers and an infra-red (IR) thermocamera to obtain an indirect measure of the currents flowing in the coils from their temperature.

5.1.1   *Measurement systems*

In order to evaluate the response of the mechanical structures to the vibrational control actions, it is necessary to measure the displacement through a specific measurement system. We adopted a scheme based on a Baumer OADM 12U6460/S35A sensor, shown in Fig. 5.3, able to provide a voltage proportional to the displacement of a given target with respect to the rest position.

The device, devoted to measure small distances, is equipped with an embedded micro-controller which provides the output voltage in the range

$[0 \div 10]$V translating the measured distance. The sensor is able to sense a target placed between a minimum distance of 16mm and a maximum distance of 120mm, providing a maximum resolution of 0.002mm.

The OADM 12U6460/S35A is a photoelectric sensor based on a working principle exploiting a triangulation. It provides a laser with wavelength of 650nm whose beam is focused through a series of lenses. An array of photodiodes is used to sense the reflection of the beam on the target. If the target changes its position, the reflection occurs with a different angle which is estimated by the micro-controller on the basis of the photodiodes response. Applying a simple triangulation, the distance with respect to the original position is calculated and the output voltage updated accordingly.

Fig. 5.3   Picture of the Baumer OADM 12U6460/S35A sensor pointing at the electromechanical structure.

As concerns the indirect measurement of the current flowing in the coils, we adopted a FLIR SC650 thermocamera, shown in Fig. 5.4. It is a portable camera able to provide thermographic videos as well as photos in a temperature range from $-40°$C up to $650°$C. It is equipped with a vanadium oxide microbolometer without a cooling system that is able to explore a wavelength range between 7.5μm and 15μm providing a thermographic image with a resolution of 640×480 pixels. Video recording can be performed with a frame-rate of 30Hz which is consistent to the angular speeds of coils and allowed us to derive information on the heating of each coil as a consequence of the current flowing through them [Kaplan (2007)].

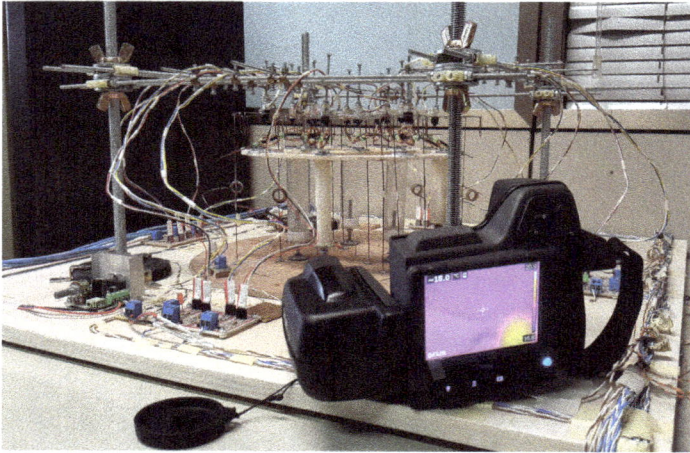

Fig. 5.4   Picture of the FLIR SC650 thermocamera.

5.1.2   *Generation of the control signals*

In Sect. 2.3, we have introduced the concept of vibrational control. In order to control the behavior of the electromechanical structures during the start-up and the regularization phases, we adopt a strategy based on an on/off controller providing a signal able to excite the imperfect dynamics of the structure.

Depending on the control task [Gawronski (2006)], the controller generates a signal which induces an oscillation on the mechanical structure. In particular, as concerns the start-up phase a high level control signal is provided to the structure according to the output of the optical sensors monitoring the angular speed of the coils. The output of the sensor array is acquired and processed through a STM32 microcontroller unit Nucleo [ST-Microelectronics (2014)] which implements the on/off switch of the control action. The high level control signal is fed to the structure if coils located at the extreme positions are not actually rotating, otherwise the high level control is turned off and the low level control is activated in order to proceed to the regularization phase.

The vibrational control signals adopted in our experiments on the electromechanical structures are signals with a broadband spectrum, so that the different modes linked to the imperfect dynamics can be continuously stimulated. The possibility to use a noisy signal, which presents the required spectral properties, has been investigated, however it reveals to be insufficient.

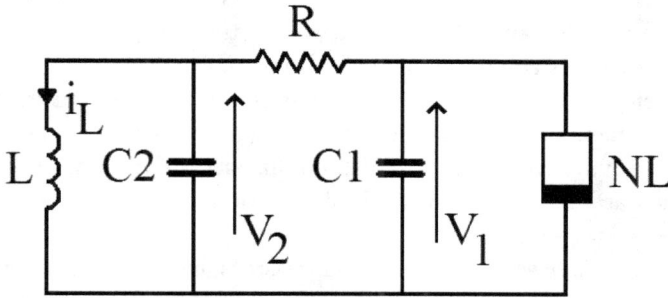

Fig. 5.5   The Chua's circuit.

After an intensive analysis of the systems behavior using periodic or noise signals to actuate the structures, whose results will be discussed later on, we retrieved that performance of the control action is sensibly increased by using chaotic signals to control the electromechanical structures. Chaos can be easily obtained from nonlinear dynamical circuits using simple and cheap off-the-shelf components. Let us consider the paradigmatic example of the Chua's circuit.

The circuit, whose original scheme is reported in Fig. 5.5, was introduced by Leon O. Chua and it is known as the first electronic circuit that was intentionally designed to generate chaotic behavior [Madan (1993); Fortuna *et al.* (2009)].

The equations of the circuit are:

$$\begin{aligned}
\frac{dv_1}{dt} &= \frac{1}{C_1}[G(v_2 - v_1) - g(v_1)] \\
\frac{dv_2}{dt} &= \frac{1}{C_2}[G(v_1 - v_2) + i_L] \\
\frac{di_L}{dt} &= -\frac{1}{L}v_2
\end{aligned} \tag{5.1}$$

where $i = g(v_1)$ is the $i-v$ characteristic of a nonlinear resistor. Performing a normalization according to $x = v_1/E_1$, $y = v_2/E_1$, $z = i_L R/E_1$, $\tau = t/RC_2$, $\alpha = C_2/C_1$, and $\beta = C_2 R^2/L$, the following rescaled equations can be written:

$$\begin{aligned}
\dot{x} &= \alpha[y - h(x)] \\
\dot{y} &= x - y + z \\
\dot{z} &= -\beta y
\end{aligned} \tag{5.2}$$

where now $h(x)$ represents the only nonlinearity in the circuit. The nonlinearity is usually expressed as a piecewise-linear (PWL) function

$h(x) = m_1 x + 0.5(m_0 - m_1)(|x + 1| - |x - 1|)$. Note that, beyond the nonlinearity often assumed with fixed parameters, only two parameters appear in the Chua's equations, namely α and β.

In order to have a reliable inductor-less circuit, we adopted the SC-CNN based implementation of the Chua's circuit dynamics, as modeled by the dynamical equations (5.2) with PWL nonlinearity. For the implementation we fix the parameters so that to obtain the double scroll Chua's attractor: $m_0 = -\frac{1}{7}$, $m_1 = \frac{2}{7}$, $\alpha = 9$, and $\beta = 14.286$.

The circuit is based on off-the-shelf resistors, capacitors and operational amplifiers. The complete scheme is reported in Fig. 5.6.

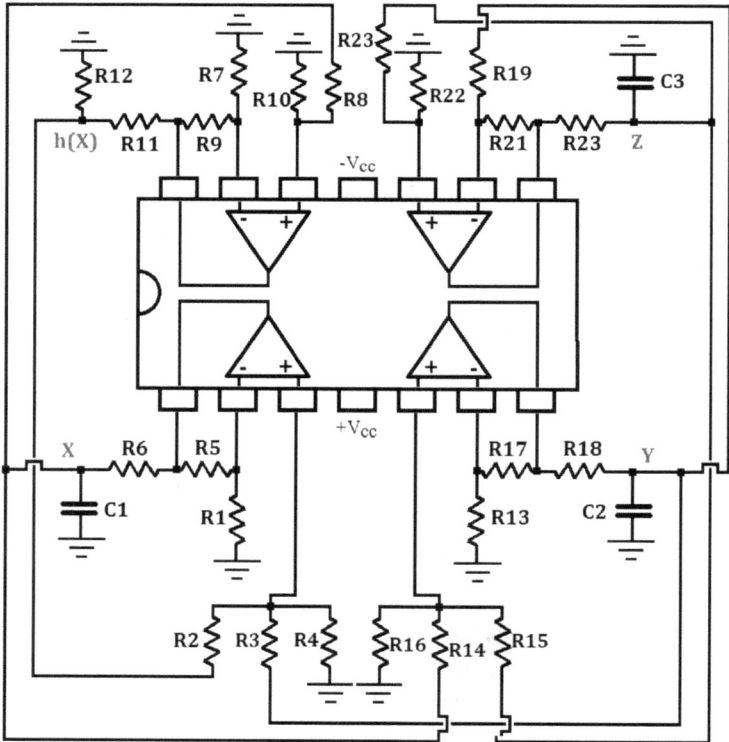

Fig. 5.6   SC-CNN based circuit implementation of the Chua's equations (5.2). Components: $R_1 = 4k\Omega$, $R_2 = 13.3k\Omega$, $R_3 = 5.6k\Omega$, $R_4 = 20k\Omega$, $R_5 = 20k\Omega$, $R_6 = 380\Omega$ (potentiometer), $R_7 = 112k\Omega$, $R_8 = 112k\Omega$, $R_9 = 1M\Omega$, $R_{10} = 1M\Omega$, $R_{11} = 8.2k\Omega$, $R_{12} = 1k\Omega$, $R_{13} = 51.1k\Omega$, $R_{14} = 100k\Omega$, $R_{15} = 100k\Omega$, $R_{16} = 100k\Omega$, $R_{17} = 100k\Omega$, $R_{18} = 1k\Omega$, $R_{19} = 8.2k\Omega$, $R_{20} = 100k\Omega$, $R_{21} = 100k\Omega$, $R_{22} = 7.8k\Omega$, $R_{23} = 1k\Omega$, $C_1 = C_2 = C_3 = 100\mu F$, $V_{cc} = 9V$.

The circuit equations associated to the implementation of Chua's circuit are the following:

$$\begin{cases} C_1 R_6 \frac{dX}{d\tau} = -X + \frac{R_5}{R_3}Y + \frac{R_5}{R_2}h \\ C_2 R_{18} \frac{dY}{d\tau} = -Y + \frac{R_{17}}{R_{14}}X + \frac{R_{17}}{R_{15}}Z \\ C_3 R_{23} \frac{dZ}{d\tau} = -Z + \frac{R_{21}}{R_{20}}Z + \frac{R_{21}}{R_{19}}Y \end{cases} \tag{5.3}$$

where:

$$h = \frac{R_{12}}{R_{11}+R_{12}} \frac{R_9}{R_8}(|X+1| - |X-1|) \tag{5.4}$$

From the match between Eqs. (5.3) and (5.2) it is possible to choose the values of the components, reported in the caption of Fig. 5.6. Furthermore, in order to obtain a spectrum whose main components are in the range of 10Hz, a temporal rescaling allowing for faster observation of the dynamical behavior is introduced. In particular, we selected a rescaling factor $\kappa = \frac{1}{C_2 R_{18}} = \frac{1}{C_3 R_{23}} = 10$.

Furthermore, choosing R_6 as a variable resistor, the different dynamical behaviors shown by the Chua's circuit by varying the single bifurcation parameter α can be observed. In fact, the parameter is related to the resistor value through the relationship $\alpha = \frac{R_5}{R_3} \frac{R_{18}}{R_6}$.

We used the state variable X, i.e. voltage across capacitor C_1, as control signal. The amplitude is varied according to the required control law through an amplifier driving the electromechanical actuators. In Fig. 5.7, we report the chaotic attractor and the oscilloscope trace of the X variable of the Chua's circuit with its power spectrum, showing the effect of the temporal rescaling factor.

5.1.3 *Actuation systems*

In order to actuate the structure with the designed vibrational control law the control signal, which is a voltage, must be converted in a mechanical displacement. Therefore, the mechanical structures have been modified adding suitably located actuators [Busch-Vishniac (2012)].

The actuation system is based on a solenoid located at points peculiar to the specific structure. The solenoid is powered by a current and interact with a permanent magnet constituting an electromagnetic actuator. It is built by 110 windings over a cylindrical support with diameter of 1cm, using 7m of isolated copper string (0.14mm diameter, 0.015mm^2 area of the section) providing a resistance of 8Ω. Building a solenoid is an easy and cheap task and allows to control the magnetic force either in voltage or in current. Moreover, the actuation system has a short response time and

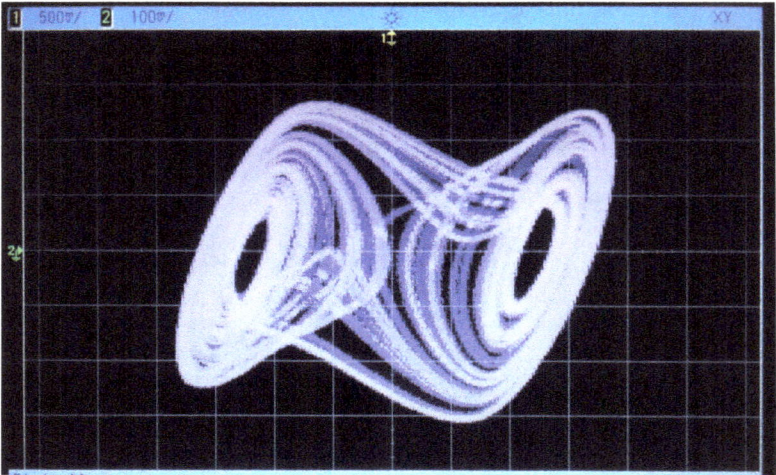

(a)

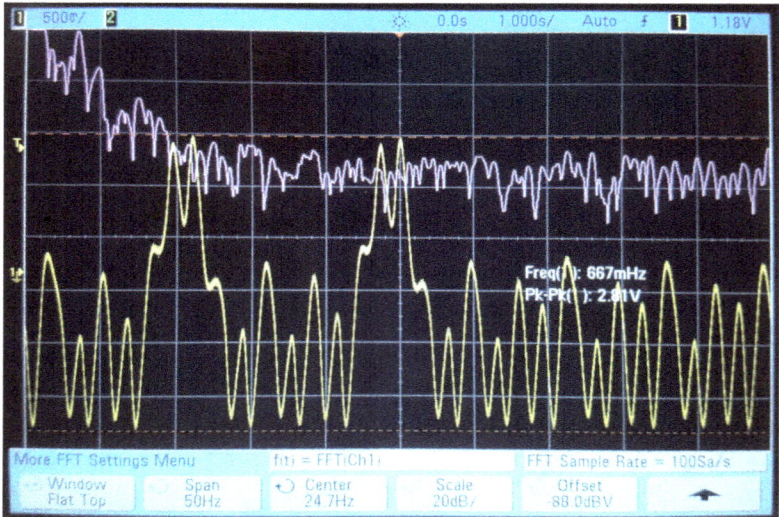

(b)

Fig. 5.7   Oscilloscope traces related to the Chua's circuit: (a) classic double-scroll attractor, (b) voltage representing the X state variable of the Chua's circuit (yellow trace) and its power spectrum (magenta trace) showing that the main components of the control signal are in the range of 10Hz.

a low impedance. The field associated to the solenoid is parallel to its axis and uniform. Its intensity and direction are proportional to the magnitude and direction of the current flowing in the windings.

The permanent magnet coupled with the solenoid is a sintered Neodymium (NdFeB) magnet with a nickel plating (Ni-Cu-Ni) of the class N45 providing a magnetic force of 2.5kg. The magnet is located on a fixed support in order to maintain its relative position with respect to the solenoid. As shown in Fig. 5.8, the magnet is located within the solenoid, ensuring to the solenoid the possibility to move and produce a displacement of the structure orthogonal to the piers.

Fig. 5.8   Detail of the actuation system: configuration of the solenoid-magnet system.

The interplay between the magnet and the solenoid depends on the current, therefore a mutual repulsive action is generated by opposite magnetic fluxes, otherwise a mutual attractive action is due to coincident magnetic fluxes. The vibration is induced on the structure alternating the direction of the current through the solenoid according to the polarity of the vibrational control signal. The power is supplied to the solenoids through an amplifier based on the TDA7297 [Skvarenina (2001); Mohan *et al.* (1988)].

The circuit, whose schematic is reported in Fig. 5.9, is a two-channel audio amplifier able to provide to an impedance of 8Ω a power up to 15W per channel, despite the fact that the control signals used have the main spectral density on a range of frequency below that for which the amplifier has been designed. We choose to adopt this scheme, whose implementation is reported in Fig. 5.10, since it is easy, reliable and low cost. The gain of the amplifier can be set to be acting on a single potentiometer.

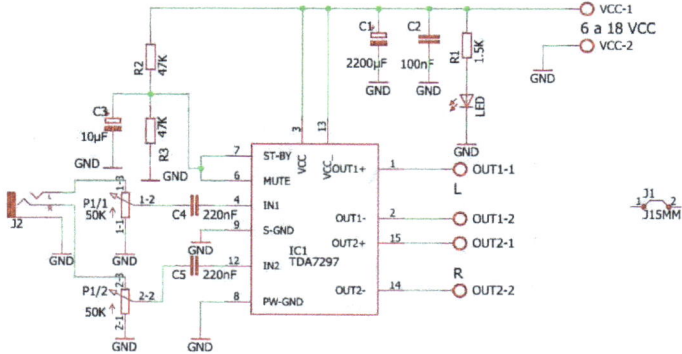

Fig. 5.9   Schematic of the amplifier driving the actuation systems.

The location of the actuation system with respect to the structure plays a crucial role on the provided effective momentum. As concerns the rectangular structure, we locate a pair of actuators, each acting on two parallel piers supporting the structure, as shown in Figs. 5.11 and 5.12. The two actuators are used for different tasks, one will provide the high level control action for the start-up phase, while the second will implement the regularizing low level action. As concerns the circular structure, the two actuators are located in correspondence to two diametrically opposed piers supporting the middle ring, as reported in Fig. 5.13. In this case the two actuators work together and in anti-phase, in order to provide a pulsating torsional momentum whose resulting vibrations excite the imperfect dynamics.

From tests carried out, we have found that the presence of the actuators on the piers of the inner circumference is not effective for the entire structure, due to the higher stiffness of inner part of the structure, and only the coils of the inner ring feel the effect. Conversely, positioning the actuators on the piers of the outer ring leads to an opposite behavior: the higher flexibility of the structure lets the solicitation to be absorbed wit-

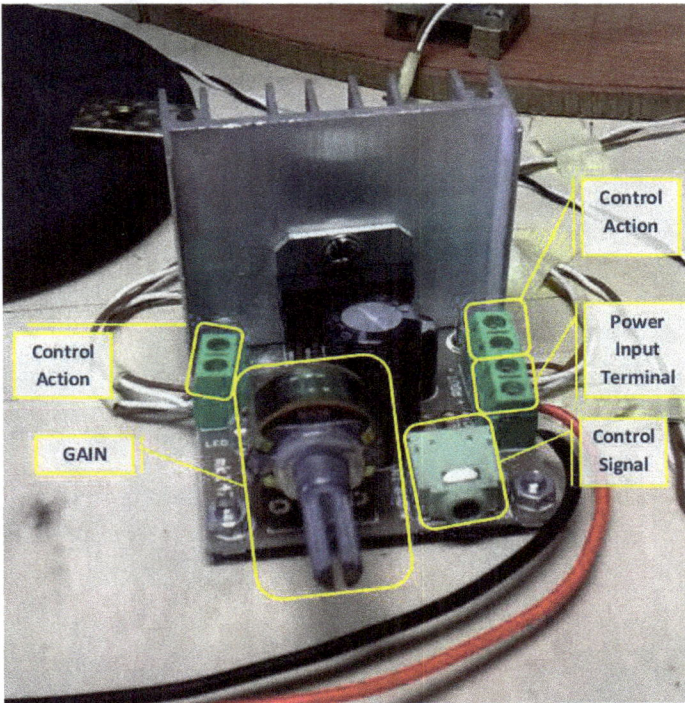

Fig. 5.10   Circuit implementation of the amplifier driving the actuation system.

hout transmitting it to the inner and central parts. In the end, the choice falls on the piers of the middle ring, positioned in correspondence with a diametrical plane, so as to exploit the central symmetry of the system.

5.2   Working Principle and Genesis of Horizontal and Vertical Vibrations

Before discussing the experimental and numerical results, in order to ex-emplify the principle generating mechanical vibrations, some snapshots, reported in Figs. 5.14(a)–(f) of a close up on the brush-pivot system are discussed.

The frames shown in Fig. 5.14 refer to a complete cycle of a single coil rotating around its pivot. They illustrate the mechanism that leads to the stimulation, in the mechanical structure, of the imperfect dynamics generating mechanical vibration both in vertical and horizontal directions.

Fig. 5.11   Location of the actuator for the start-up phase on the rectangular structure.

The effect of friction between the supporting trails and the coil is to lift it up along one of the sides of the support so that, when gravity makes the coil fall down, it hits the trail. The impact with the structure generates the vibration. The continuous rotation of each single coil produces a series of impacts which generate the horizontal and vertical vibrations to which the structures are subjected.

During the complete cycle of the single coil we can observe the trend of the angular speed by analyzing the output of the optical sensor. In Fig. 5.15 the angular speed in hertz of the coil is reported over 60s of observation. In order to obtain the curve in Fig. 5.15, the output of the optical sensor is divided in 60 bins, corresponding to each 1s of observation.

Fig. 5.12 Location of the actuator for the regularization phase on the rectangular structure.

Fig. 5.13 Location of the two anti-phase actuators for the circular structure.

(a)   (b)

(c)   (d)

(e)   (f)

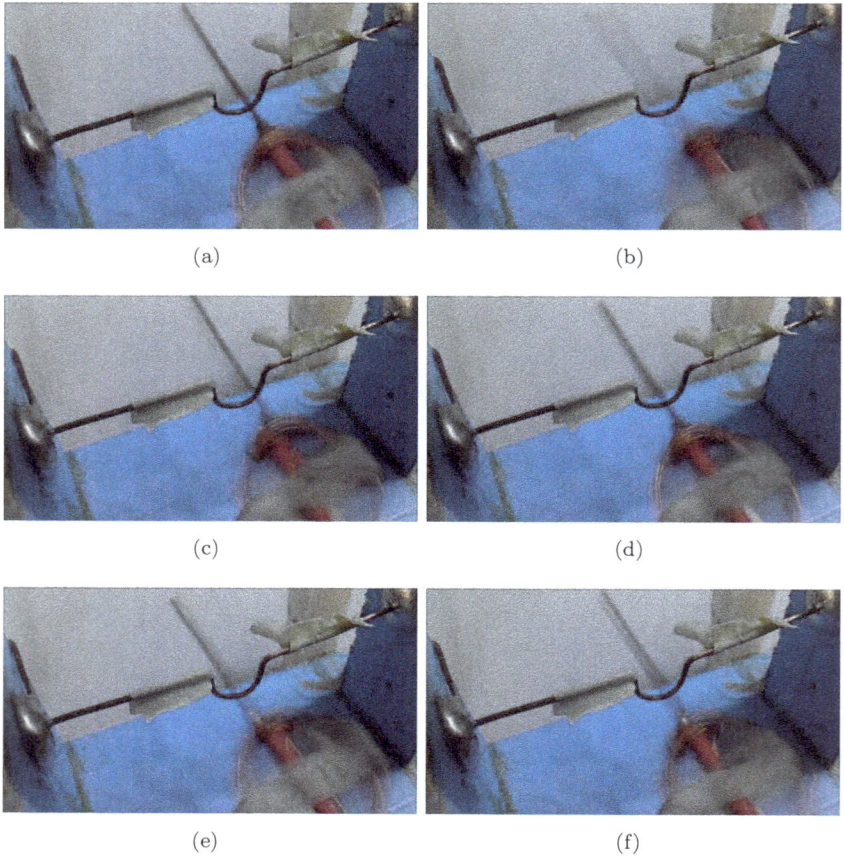

Fig. 5.14   Complete cycle of a single coil around the brush-pivot support: (a) friction causes the rotating coil to climb the left part of the support, (b) gravity makes the coil fall down impacting the structure, (c) coil tends to climb on the opposite side of the support, (d) falls down hitting again the support, (e) coil rotates and climbs the left part of the support, and (f) falls down hitting again the structure.

As discussed in Chapter 3, the angular speed is calculated as the inverse of the temporal distance between each peak of the output signal and the second one, therefore each point of the curve in Fig. 5.15 represents the average speed calculated for each bin, i.e. over 1s.

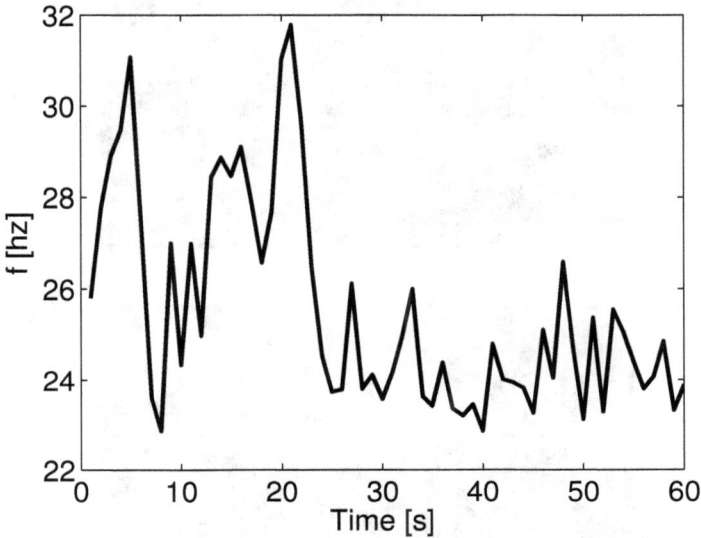

Fig. 5.15 Trend of the angular speed of a single rotating coil over 60s. Each point represents the angular speed averaged over 1s.

It is also interesting to observe the peculiar behavior of the current flowing through the single coil. In order to use a digital oscilloscope, we put in series with the positive terminal of the power supply a shunt power resistor of 1Ω and display the trend of the voltage across it. The oscilloscope traces reported in Figs. 5.16(a)–(b) show that during the rotation the coil is not conductive and, therefore, the voltage drops to zero, furthermore, during the conductive phase a series of spikes can be observed.

5.3 Experiments on the Rectangular Structure

In order to determine the structure's mechanical response to the applied forces and to define its resonance frequency, the actuators, which are positioned longitudinally, have been excited with different canonical signals at varying frequencies. The frequency response, and the corresponding coherence function, have then been derived considering as input a sinusoidal mechanical vibration and as output the displacement of the structure measured by the laser. We used a Dynamic Signal Analyzer Agilent 35670A to derive the frequency response of the structural displacement with respect to

(a)

(b)

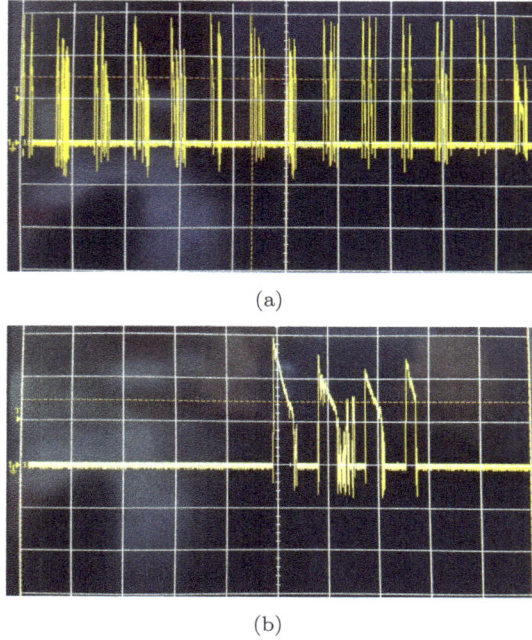

Fig. 5.16   Oscilloscope traces of the voltage across the shunt resistor: (a) series of bursting, (b) detail of a single burst. Scales: horizontal 2ms/div, vertical 1V/div

a sinusoidal input. The Dynamic Signal Analyzer Agilent 35670A allows us to also verify the quality of the reconstructed frequency response providing the evaluation of the corresponding spectral coherence $C_{xy}(f)$ defined as

$$C_{xy}(f) = \frac{|G_{xy}|^2}{G_{xx}(f)G_{yy}(f)}$$

where G_{xy} is the cross-spectral density between the input and the output, G_{xx} and G_{yy} are the auto-spectral density of the input and the output, respectively [Priemer (1991); Bendat and Piersol (2011)]. The information provided by the spectral coherence allows to asses the reliability of the frequency response estimation, where values of $C_{xy}(f)$ close to 1 indicate a high confidence level of the estimation, while values of $C_{xy}(f)$ close to zero indicate a low confidence level.

The results related to this analysis will be discussed in the following sections.

5.3.1 *Mechanical behavior*

We first focus on the response to period vibrations on the structure at different frequencies. This has been done in order to highlight possible filtering effects according to the system. The signal was sent to the electromagnetic actuators via a power device capable to provide the required current. The tests were carried out by maintaining the amplitude of all signals fixed, while frequencies are chosen within a range of values significant for the mechanical behavior of the structure. The system response in terms of displacements of piers, in particular those located in the opposite position with respect to the solicited ones, was detected using the laser that returns a voltage proportional to the displacement with respect to the equilibrium (zero position) of the measuring point. All signals were sent to a digital oscilloscope that allowed us to view them, make a first qualitative estimate and calculate the power spectrum of the response signal in order to determine its harmonic components.

The obtained results are shown in Figs. 5.17–5.19.

The analysis of the obtained results clarifies how the structure has a resonant frequency centered around 4Hz. To further investigate this point, we can drive the actuation system with a sinusoidal signal whose frequency sweeps up to 50Hz, and use the Dynamic Signal Analyzer to derive the frequency response and the spectral coherence. The results are reported in Fig. 5.20 and show that the frequency response displays a peak around 4Hz in correspondence to a high spectral coherence, thus confirming our previous results. Furthermore, the results provided by the Dynamic Signal Analyzer allow us to discover that the frequency response shows a broad spectrum around the resonance frequency, instead of a single peak. This confirms again that the use of vibrational control in a chaotic signal with high spectral density in the same range is worth investigating.

5.3.2 *Start-up phase and angular speeds characterization*

Let us now focus on the start-up phase, whose experimentation is represented in the frames reported in Figs. 5.21(a)–(c). Consistently to what was observed in the mathematical model, when the voltage is supplied to the coils, only few of them start to rotate since the magnetic torque is not enough to move them from the rest position. In this condition, when the control action is turned on, the structure is subjected to a strong broadband mechanical vibration, driven by the Chua's circuit variable, and induced by the first electromagnetic actuator which elicits the imperfect dynamics of

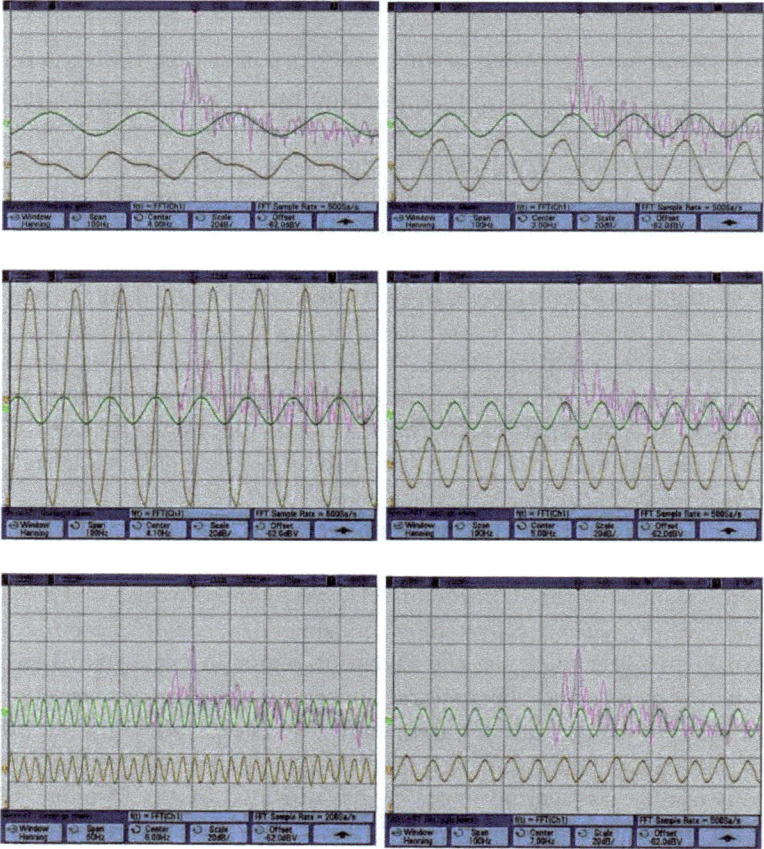

Fig. 5.17 Displacement (orange trace) produced by vibrations generated by a sinusoidal signal (green trace) with frequency f, and related power spectrum (magenta trace) as evaluated by a digital oscilloscope: (a) $f = 2$Hz, (b) $f = 3$Hz, (c) $f = 4$Hz, (d) $f = 5$Hz, (e) $f = 6$Hz, and (f) $f = 7$Hz.

the structure and allows all the coils to start rotating.

The measurements reported in Fig. 5.22 cover a period of $60s$ during which the system is powered, at $t = 0s$, and the start-up control action is turned on at $t = 5s$. They show the angular speed of each coil averaged over a time window of $1s$, and color coded according to the reported color bar. At the beginning, the coils tend to remain in the rest position until the control action is activated. After about $30s$, all coils are rotating with different angular speeds, and the start-up phase is completed.

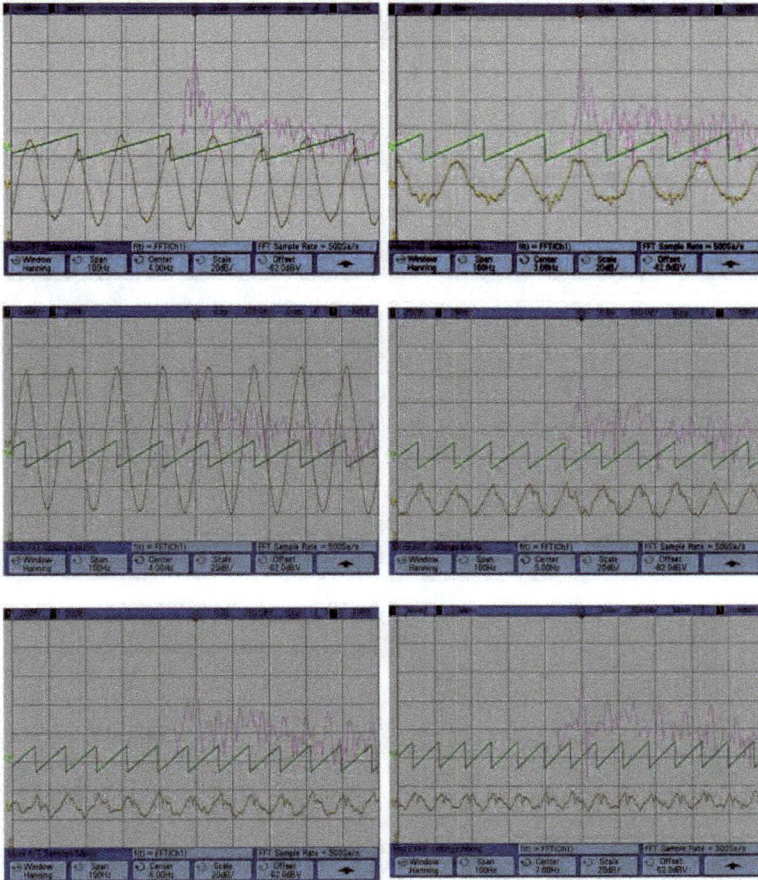

Fig. 5.18 Displacement (orange trace) produced by vibrations generated by a triangular-wave signal (green trace) with frequency f, and related power spectrum (magenta trace) as evaluated by a digital oscilloscope: (a) $f = 2$Hz, (b) $f = 3$Hz, (c) $f = 4$Hz, (d) $f = 5$Hz, (e) $f = 6$Hz, and (f) $f = 7$Hz.

The comparison between the experimental and numerical analysis allows to assess that the model is able to reproduce the steady-state behavior encompassing the diversity of friction factors between coils. Considering an uncertainty of 10% among the coils' friction factors K_i in the model, it is possible to obtain a distribution of angular speeds similar to what was observed in the experiment.

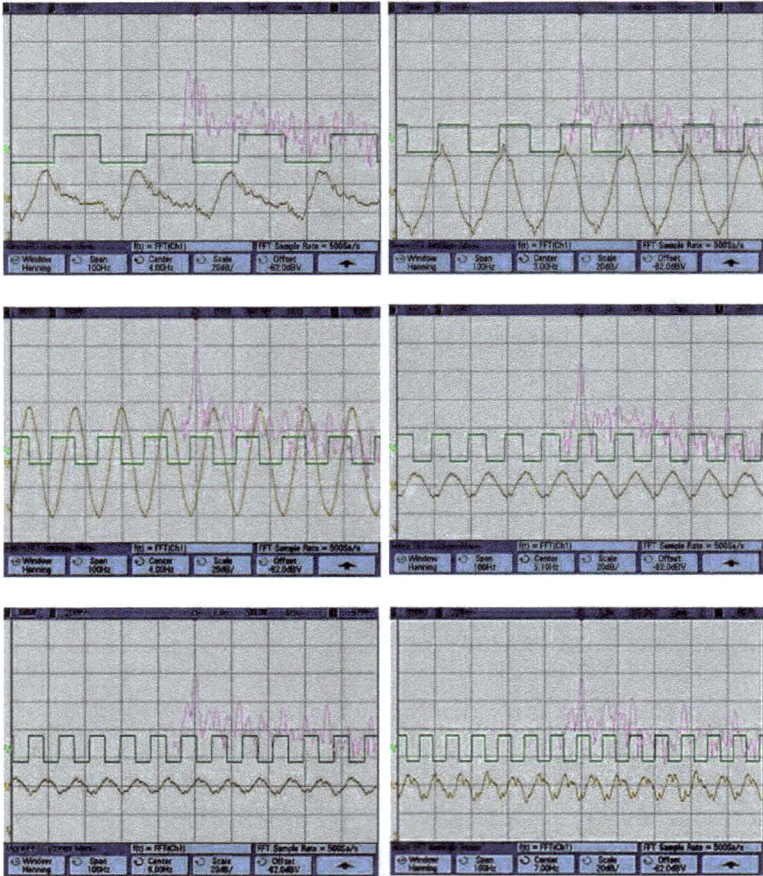

Fig. 5.19 Displacement (orange trace) produced by vibrations generated by a square-wave signal (green trace) with frequency f, and related power spectrum (magenta trace) as evaluated by a digital oscilloscope: (a) $f = 2$Hz, (b) $f = 3$Hz, (c) $f = 4$Hz, (d) $f = 5$Hz, (e) $f = 6$Hz, and (f) $f = 7$Hz.

This ensures that the transition from the rest state to the rotation takes place with the minimum energy supply by the devices external to the structure, since, for the principle of resonance, it will respond effectively to the solicitation.

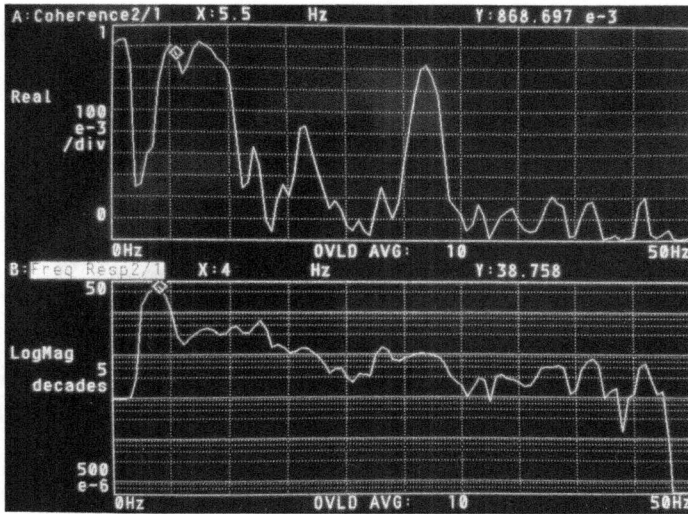

Fig. 5.20 Experimental estimation of the frequency response of the rectangular structure with respect to induced vibrations. Displacements are measured by means of the laser sensor located on one supporting pier. Upper panel: spectral coherence $C_{xy}(f)$; lower panel: frequency response considering as input a sinusoidal waveform used to drive the actuation system and as output the displacement of the rectangular structure measured by the laser sensor.

5.3.3 *Regularizing the angular speeds: weak vibrational control*

As concerns the regularization of the coils' angular speeds, we provided a further active vibration to the structure which is characterized by a lower amplitude. The performance of this further control action, in fact, is strongly related to its spectrum properties. In order to characterize the role of the regularizing control signal, the coil angular speeds have been evaluated averaging them in time over the steady-state behavior and calculating the standard deviation σ_S of the ten different coil angular speeds. The curve reported in Fig. 5.24 represents σ_S as a function of the main frequency of the transversal vibration induced by the second control action. Corresponding to a main frequency of 30Hz, a minimum can be observed: the angular speeds are more homogeneous when the structure is excited at a frequency comparable to the mean rotating speed of the ten coils.

(a)

(b)

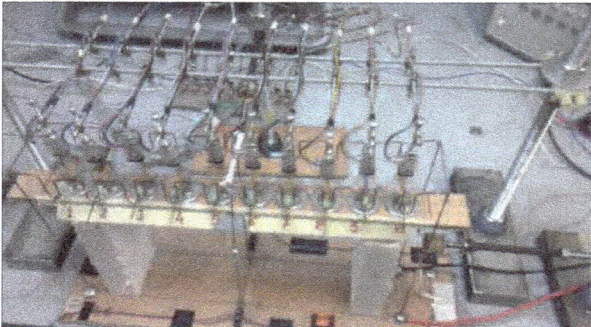

(c)

Fig. 5.21 Sequence of the experiment: (a) voltage is supplied to coils but only few start rotating, (b) the first control action is turned on until (c) all the coils are actually rotating.

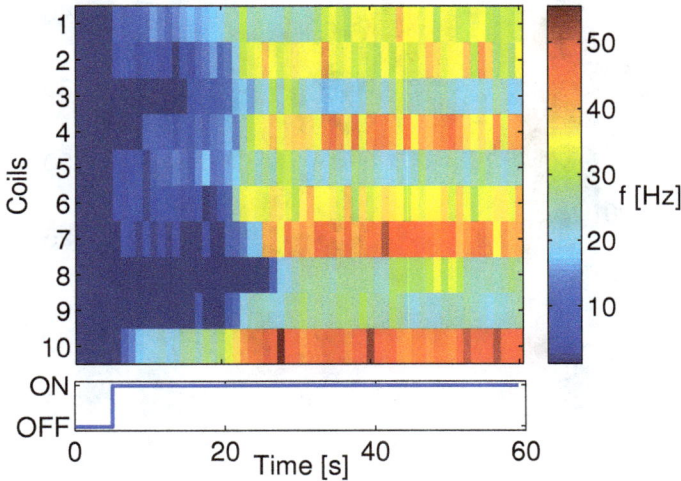

Fig. 5.22 Angular speeds (Hz) of the ten coils during the start-up phase averaged over a time window of 1s. The start-up control action is activated at $t = 5s$.

The complete experiment from start-up to regularization is described in Fig. 5.25. The time evolution of angular speeds of the ten coils are reported averaging each of them over a time window of 1s. The on/off states of the two control actions determine the behavior of the system: the start-up control action is turned on at $t = 5$s when almost all coils are unable to start rotating. After a brief transient, all coils are actually rotating but their angular speeds are significantly different. The regularizing control action, orthogonal to the length of the structure, is then turned ON at $t = 22$s inducing a significant alignment of the angular speeds.

The effect of the regularization action is also shown from the patterns obtained by using the infrared thermocamera FLIR sc450 taken with a resolution of 100dpi and an exposure time of 20ms. The snapshots reported in Fig. 5.26 show the comparison between the two cases in which the transversal forcing is off or on. When a coil rotates at a lower speed, the current flowing through it is increased and consequently its temperature. In Fig. 5.26(a) the regularizing control action is turned off and several coils are hotter than the other indicating that the speeds are not homogenous, when the control action in turned on (Fig. 5.26(b)) all the coils have almost the same temperature. The sharper appearance of three coils in Fig. 5.26(a) is due to the fact that when the angular speed of a coil is reduced, a high

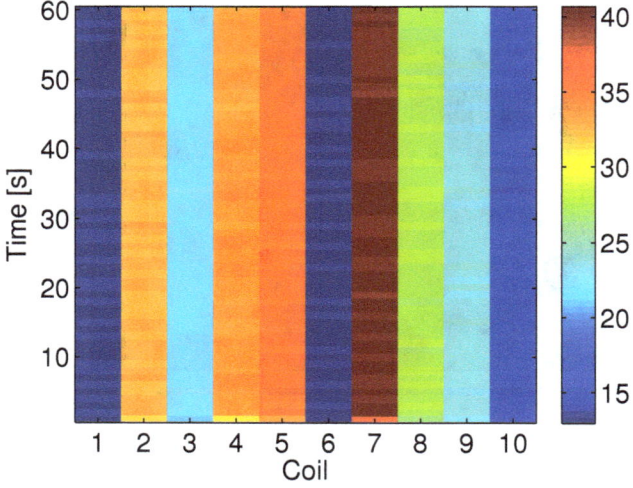

(a)

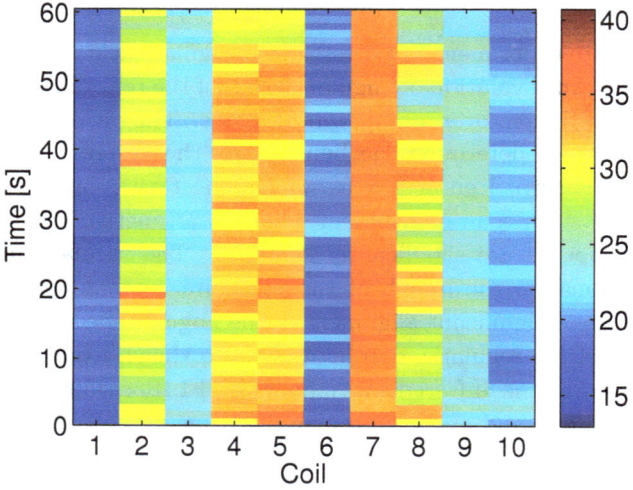

(b)

Fig. 5.23 Angular speed (Hz) of the ten coils: (a) experimental steady-state after star-tup; (b) numerical simulation introducing a 10% uncertainty on the friction value.

current flows through rapidly heating it and thus creating an evident difference with room temperature.

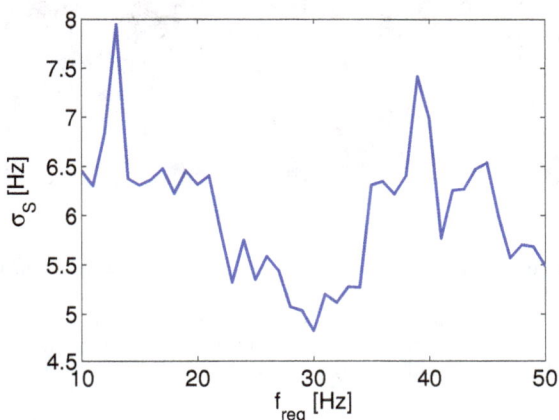

Fig. 5.24 Standard deviation σ_S of the angular speeds as a function of the frequency of the regularizing control signal f_{reg}. In correspondence of the minimum, coils rotate at almost the same frequency.

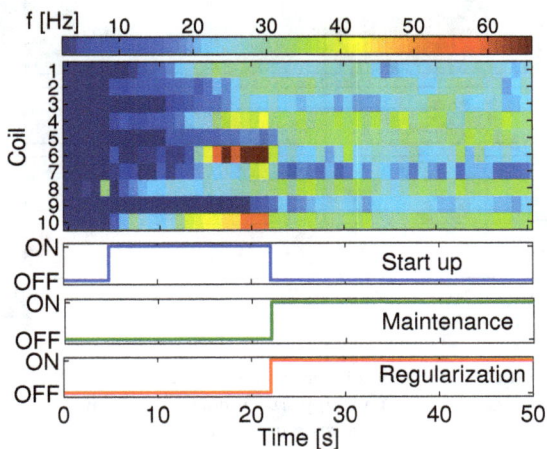

Fig. 5.25 Upper panel: angular speeds of the 10 coils averaged over a time window of 1s. Lower panels: (from top to bottom) ON/OFF status of the start-up (high) control signal, of the maintenance (low) control signal, and of the regularizing control signal.

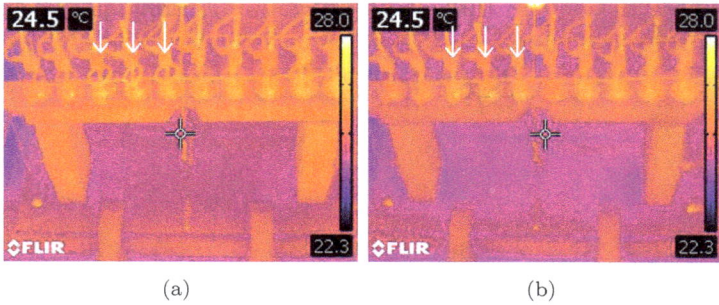

(a)   (b)

Fig. 5.26   Infrared camera acquisition of the mechanical system when the regularizing loop is (a) off and (b) on. Coils rotating at lower speeds are characterized by higher temperatures. The comparison between parts (a) and (b) of the figure shows that three coils, highlighted with white arrows, rotate with very slow speed in (a), while with speeds similar to that of the other coils in (b).

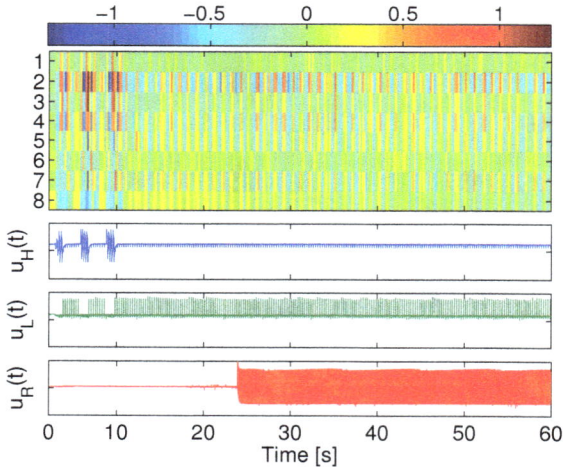

Fig. 5.27   Upper panel: amplitude of horizontal displacements Δh_i of each segment of the structure connecting adjacent coil supports in normalized units. Lower panels: (from top to bottom) ON/OFF status of the startup (high) control signal, of the maintenance (low) control signal, and of the regularizing control signal.

The system behavior during the complete experiment is now investigated by monitoring the horizontal displacement of each of the eight segments of the mechanical structure connecting two adjacent coil supports. In Fig. 5.27 the normalized displacements Δh_i, acquired through an array of TCRT5000 reflective optical sensor are reported for the eight segments. The three bottom plots show the on/off status of the control signals, namely the start-up high level control and the low-level regularizing control. The behavior of each segment is controlled by the control signals: during the start-up, phase segments are subjected to oscillations of higher amplitude, while during the regularizing phase all segments tend to oscillate with similar amplitudes.

Clearly, a complete synchronization of the coil angular speeds, i.e. a scenario where each coil rotates at the exact same frequency, is a not realistic task [Nijmeijer and Rodriguez-Angeles (2003)]. However, the effect of the regularizing control action is evident by measuring the spectral synchronization of the coil angular speeds. For spectral synchronization we mean the condition in which the same frequency spectrum can be retrieved between two or more signals. Let us consider the oscilloscope traces reported in Fig. 5.28. They represent the power spectrum of the output of the optical sensors monitoring the angular speeds for four different coils. The acquisition has been realized after the start-up phase is completed but intentionally turn off the regularizing control action. The four power spectra are not superposed, indicating that the four coils are rotating around a proper main frequency. The introduction of the low level vibrational control induces an evident modification of the spectral properties of the four angular speeds, as shown in Fig. 5.29. Indeed, the effect of the control action can be appreciated by the increase of spectral components around 4Hz and up to 10Hz, which are the main bands of the chaotic signal used to drive the actuation systems. Furthermore, the whole spectra tend to align, showing a sensibly increased regular behavior with respect to that reported in Fig. 5.28.

It should be further noticed that the level of the control signal must be suitably tuned. A high-level produces oscillations too heavy for the regularization, while a very low-level is filtered by the structure. An optimal value of the control signal amplitude has to be experimentally retrieved to enhance the spectral synchronization. Therefore, this is a phenomenon similar to the stochastic resonance [Andò and Graziani (2012)] but it differs in the sense that stochastic resonance is appreciated specifically in time domain, while spectral synchronization is appreciated in frequency domain.

This effect is essentially due to the existence of the imperfect dynamics that interact with the uncertainty regularizing the dynamical behavior.

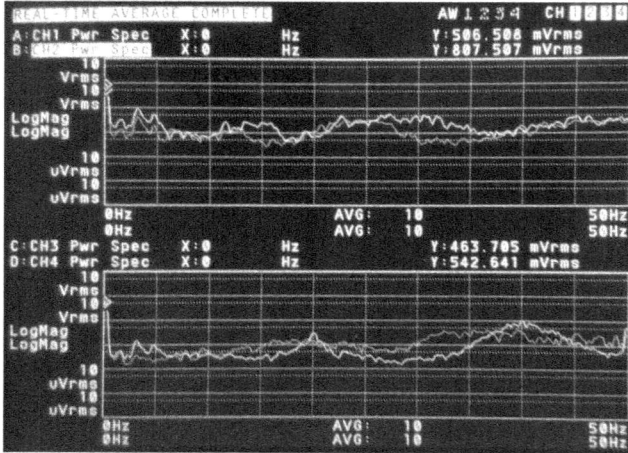

Fig. 5.28 Spectral analysis of the angular speeds of four coils rotating over the rectangular structure, when the low level control action is off.

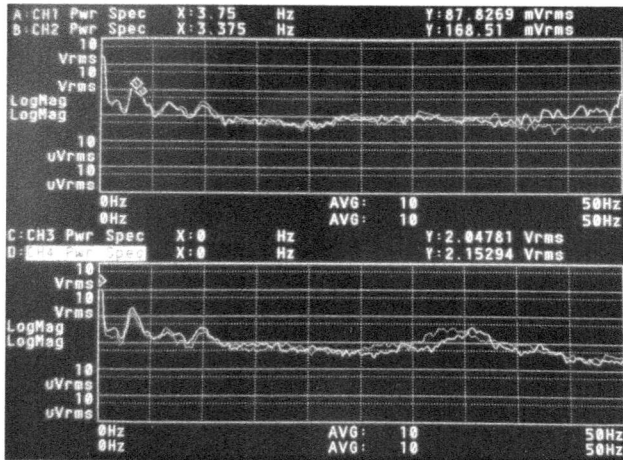

Fig. 5.29 Spectral analysis of the angular speeds of four coils rotating over the rectangular structure, when the low level control action is on.

5.4 Experiments on the Circular Structure

A study of the mechanical behavior, similar to that seen for the linear structure, was conducted on the circular structure, by inducing vibrations through the actuation system.

5.4.1 *Mechanical behavior*

Given the particular shape of the structure and its specific spatial organization, the study has taken into account not only the frequency with which the structure itself is mechanically stressed, but also the direction that the forces applied by the actuators must have, in order to obtain the best response from the structure as a whole.

The mode excitation of the electromechanical actuators is also optimized. In the first experiment the structure is stressed by means of only one of the actuators taken into account. As can be seen from the obtained data, the response of the system is not very effective both in terms of response to stress and in terms of start-up of the coils. The data obtained are shown in Figs. 5.30–5.32. Ultimately, in the light of the obtained data, the possibility of controlling the structure with only one actuator is ineffective and discarded.

We now focus on the possibility to solicit the structure with two actuators. The excitation signal is the same for both actuators to ensure consistency between the two sources in terms of amplitude, frequency and phase. The two actuators can be activated in two different ways on the basis of the direction of the applied forces at the points of stress. It is assumed that the two actuators have an in-phase behavior, when the vectors of the applied forces in the piers have the same direction with respect to their plane. We say that the two actuators have an anti-phase behavior, when the two vectors of the forces applied on the abutments are anti-parallel. The different mode of excitation of the two actuators results in a different mechanical behavior of the structure. In the first case, the whole structure is forced to oscillate along the vertical plane of the active piers. The data obtained are shown in Figs. 5.33–5.35. In the second case, the central ring of the structure is affected by a twisting motion, with respect to the central axis, which leads the whole structure to rotate at an angle proportional to the displacement applied by the actuators. It is this configuration, among all, the one that gives the best results in terms of mechanical stress of the structure and start-up of the coils. The data obtained are shown in Figs. 5.36–5.38.

The anti-phase configuration has been also used to identify the frequency response of the circular structure by using the Dynamic Signal Analyzer. The displacement of a pier has been considered as the output corresponding to a sinusoidal sweep signal provided to the two actuators. The spectral coherence has been also evaluated. Results are shown in Fig. 5.39, from which a peak in the frequency response can be retrieved around 7Hz, a result that confirms the preceding observations and lead us to use a chaotic signal whose band mainly lies within the same range.

5.4.2   *Start-up phase and angular speeds characterization*

Let us now discuss the pattern of the angular speeds of the coils in the circular structure. The measurements reported cover a period of 30s of the experiment from startup and show the variation of the average angular speed of each coil (the average is performed on a time window of 1s). The first problem faced is the choice of the type of stress applied to the system, or the manner in which actuators are excited, in order to ensure that, after a certain finite time interval, all the coils are rotating. In addition, this has to be realized with the minimum input of energy from the power supply that provides the electrical power to the system itself.

The series of tests carried out on the system has been performed with the same methodology used for the determination of the resonance frequency of the system. The actuators are energized in different ways and at different frequencies; the timing of implementation of the coils is detected and shows which applied signals and which manner of excitation provide the best response in terms of activation times. The signal generated by the infrared sensor, which detects the speed of the coil, is evaluated, on average, in a time window of 1s. Its value is encoded in a chromatic scale that returns, in qualitative terms, the average speed of each coil of the system in the various temporal windows, for the duration of the experiment that lasts 30 seconds. This test will be positive if all the coils start before the time window of the experiment elapses. The obtained results are shown in the following figures.

As far as the excitation with a single actuator is concerned, only one result at 3Hz is reported, because the structure has not responded in a satisfactory manner. It is clearly visible from Fig. 5.40 how not all the coils are started in the temporal observation window. Some columns are completely blue and this shows that the corresponding coils are not rotating (namely coil 5, 8, 10, 12, 13, 14, 16).

From the analysis of the previous graphs, we conclude that the best condition for the start of the system occurs at the frequency of 6 Hz: this occurs immediately before the resonance frequency. All the coils are started in a relatively short time interval and, while keeping the control active, have certain uniformity in the average speed of rotation.

5.4.3 Regularizing the angular speeds: weak vibrational control

The circular structure has been connected to a feedback control system, which is able to start the coils and keep them rotating. The experimental results have been summarized in Fig. 5.47.

In this figure, it can be noticed that all the coils are working as $t > 21$s. From their colors, uniformity in speed can be deducted. The analysis of the thermographic images reported in Fig. 5.48 provides also a further insight. As it is possible to observe, when the regularizing action is turned off, coils in different areas of the circular structure are rotating at different angular speeds and, therefore, the current flowing through them is different. In Figs. 5.48(a)–(b), some coils are hotter, denoting that their angular speeds are lower. Turning the control action on leads to the situation shown in Figs. 5.48(c)–(d), where all coils have the same color, indicating similar temperatures and, indirectly, similar angular speeds.

To further clarify this we can observe the effect of the regularization control action on the power spectral properties of four coils located in far areas of the circular structure. In Fig. 5.49 the structure is not subjected to the control action and the coils are rotating around different angular speeds. The effect of the regularization is evident from the power spectra reported in Fig. 5.50.

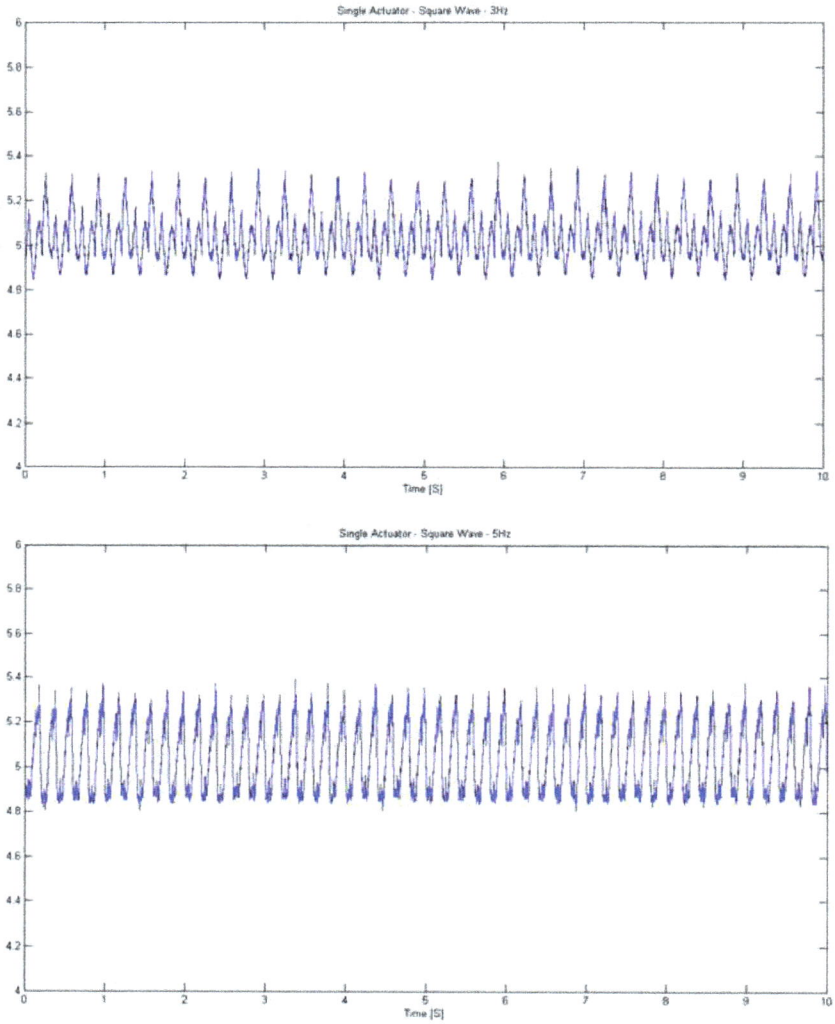

Fig. 5.30   Displacement of the internal pier of the circular structure excited by a single electromechanical actuator driven by a sinusoidal waveform of different frequencies: (a) 3Hz, (b) 5Hz.

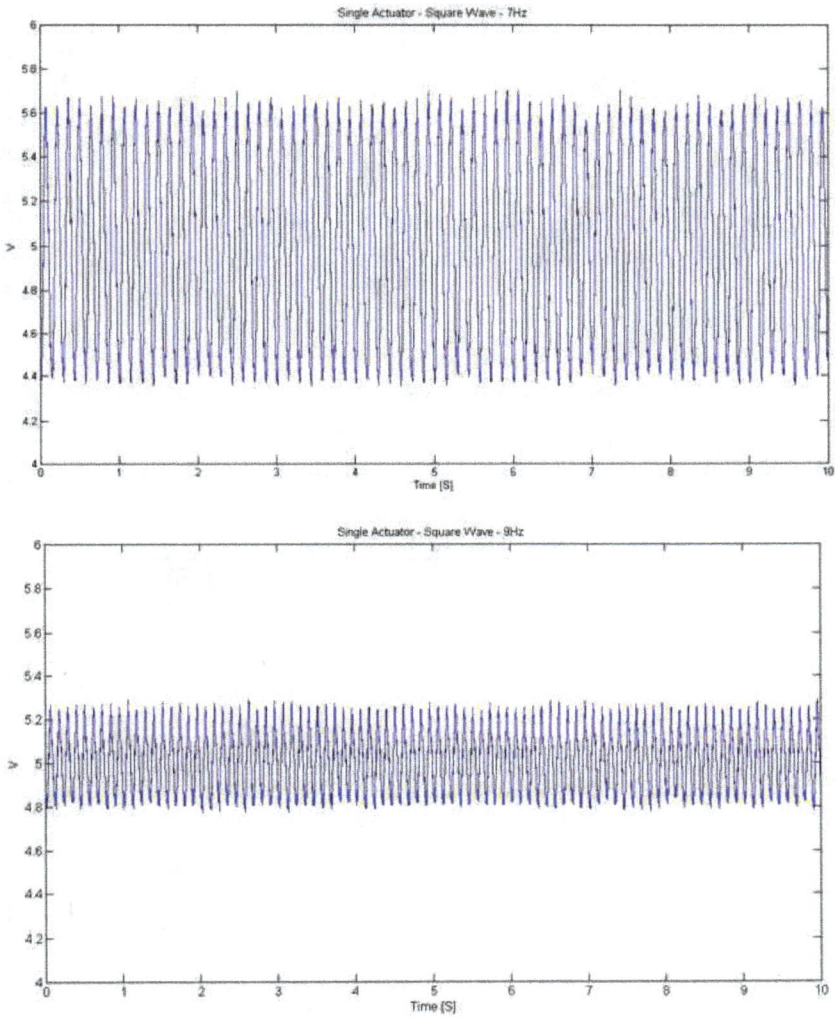

Fig. 5.31 Displacement of the internal pier of the circular structure excited by a single electromechanical actuator driven by a sinusoidal waveform of different frequencies: (a) 7Hz, (b) 9Hz.

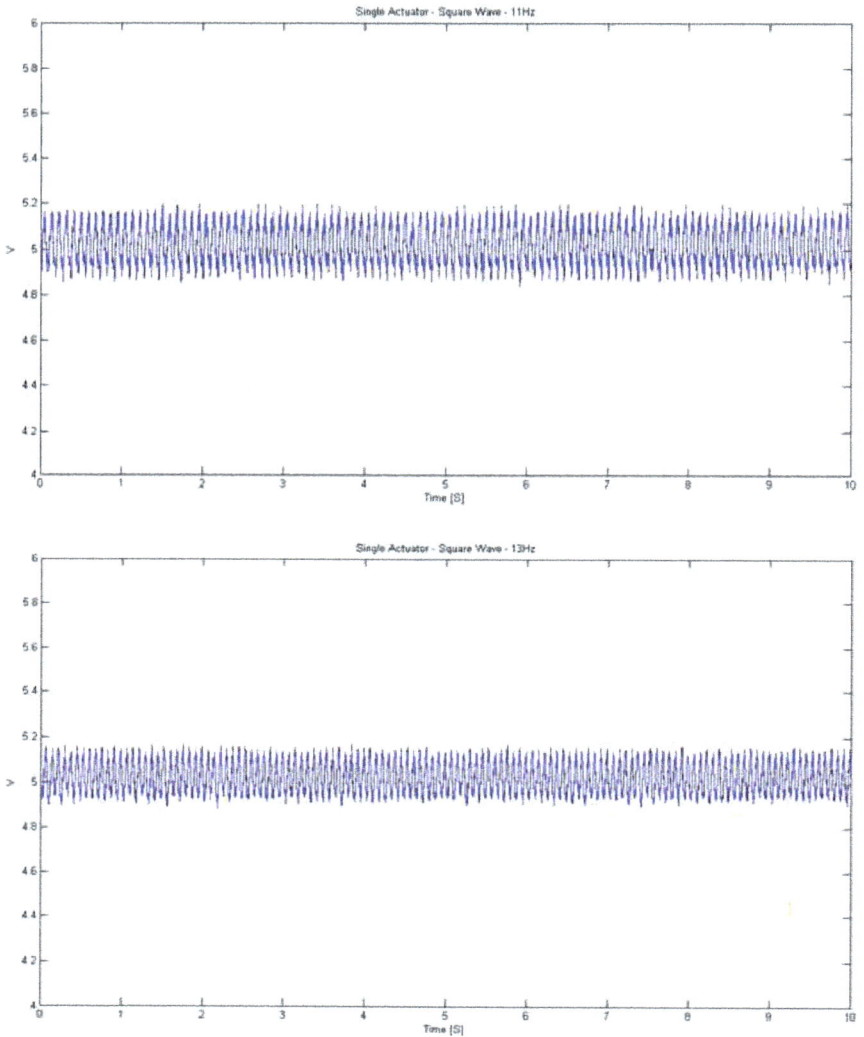

Fig. 5.32   Displacement of the internal pier of the circular structure excited by a single electromechanical actuator driven by a sinusoidal waveform of different frequencies: (a) 11Hz, (b) 13Hz.

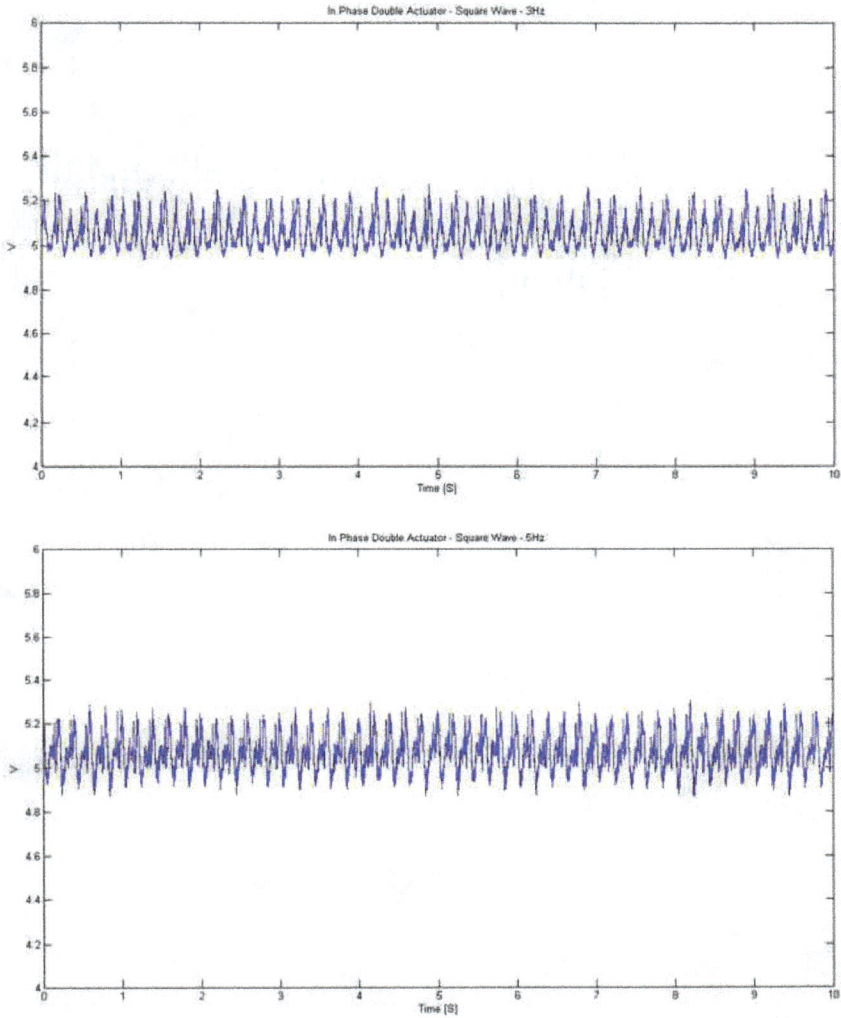

Fig. 5.33 Displacement of the internal pier of the circular structure excited by a pair of in-phase electromechanical actuators driven by a sinusoidal waveform with different frequencies: (a) 3Hz, (b) 5Hz, (c) 7Hz, (d) 9Hz, (e) 11Hz, (f) 13Hz.

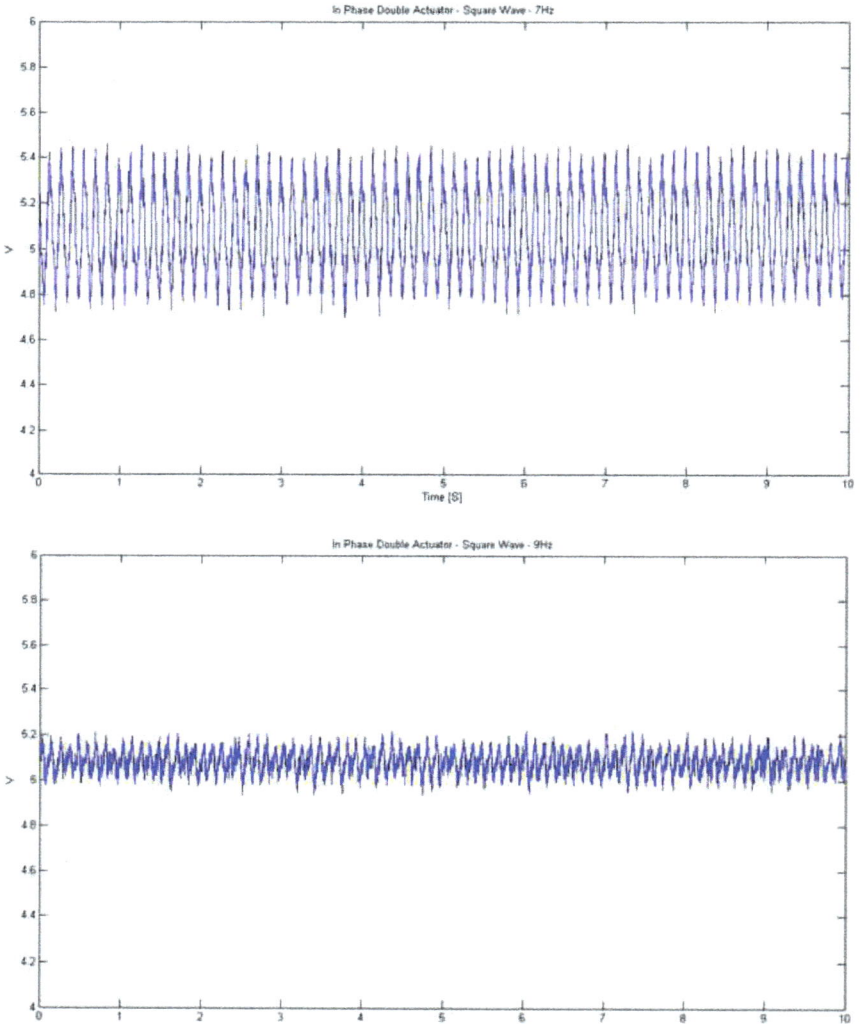

Fig. 5.34 Displacement of the internal pier of the circular structure excited by a pair of in-phase electromechanical actuators driven by a sinusoidal waveform with different frequencies: (a) 3Hz, (b) 5Hz, (c) 7Hz, (d) 9Hz, (e) 11Hz, (f) 13Hz.

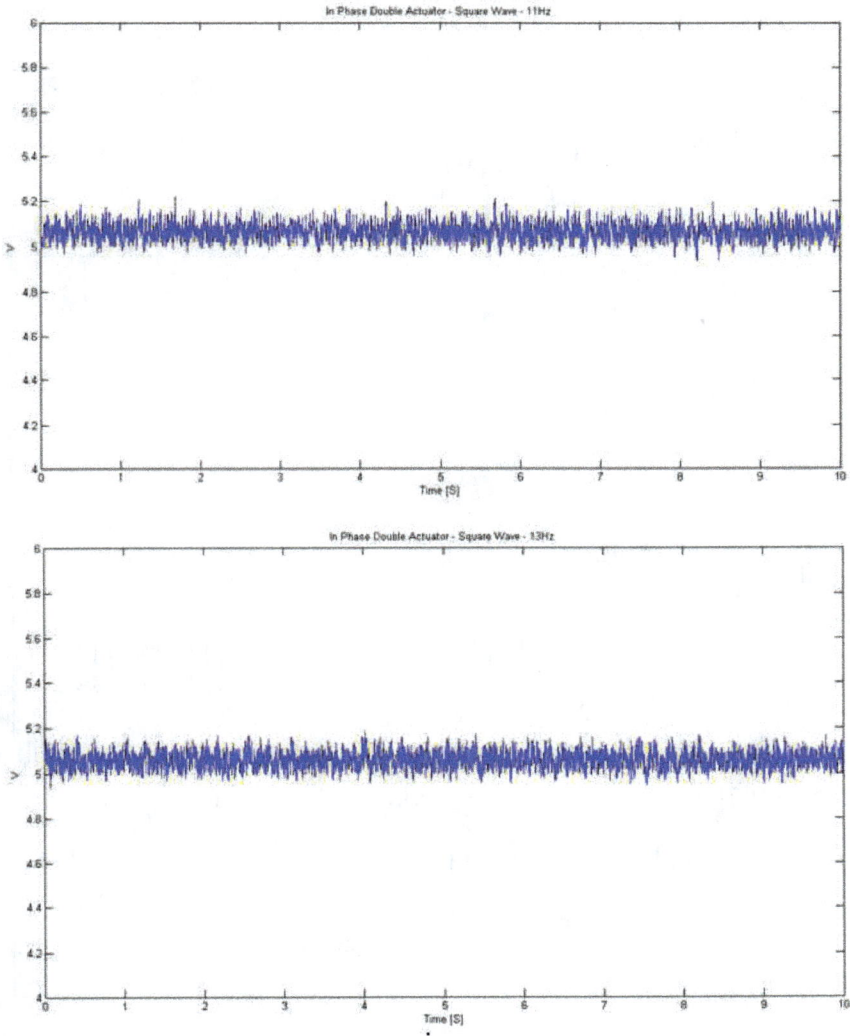

Fig. 5.35 Displacement of the internal pier of the circular structure excited by a pair of in-phase electromechanical actuators driven by a sinusoidal waveform with different frequencies: (a) 3Hz, (b) 5Hz, (c) 7Hz, (d) 9Hz, (e) 11Hz, (f) 13Hz.

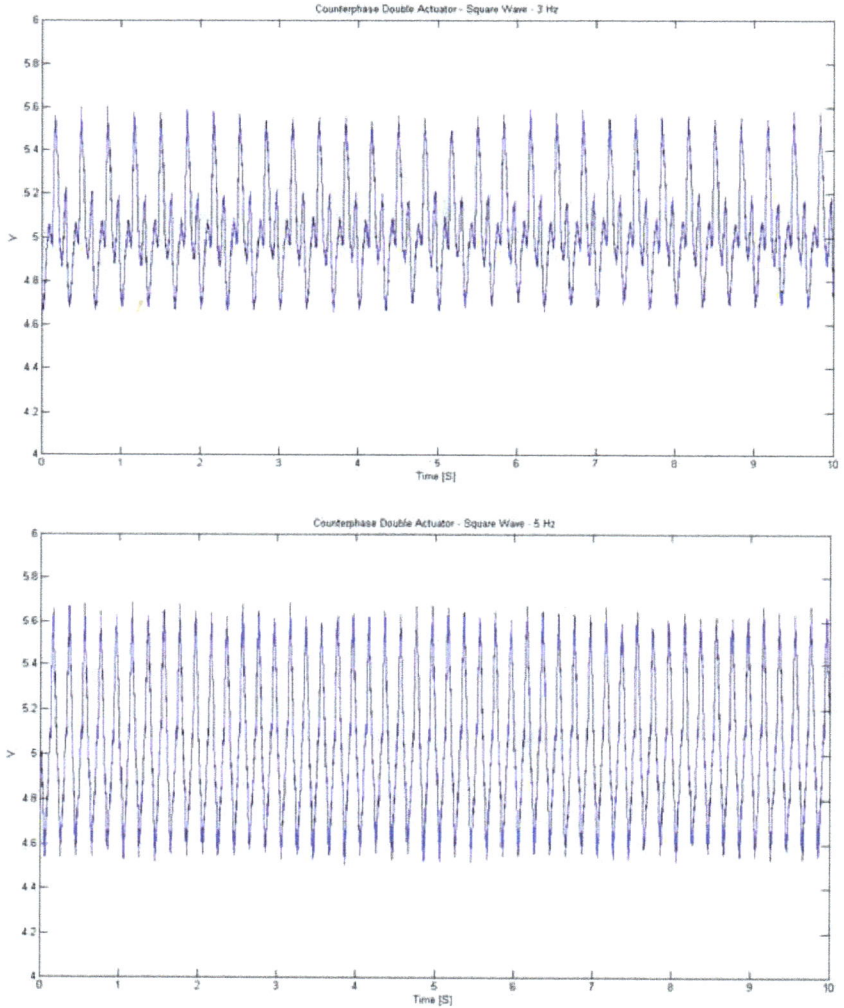

Fig. 5.36   Displacement of the internal pier of the circular structure excited by a pair of anti-phase electromechanical actuator driven by a sinusoidal waveform with different frequencies: (a) 3Hz, (b) 5Hz, (c) 7Hz, (d) 9Hz, (e) 11Hz, (f) 13Hz.

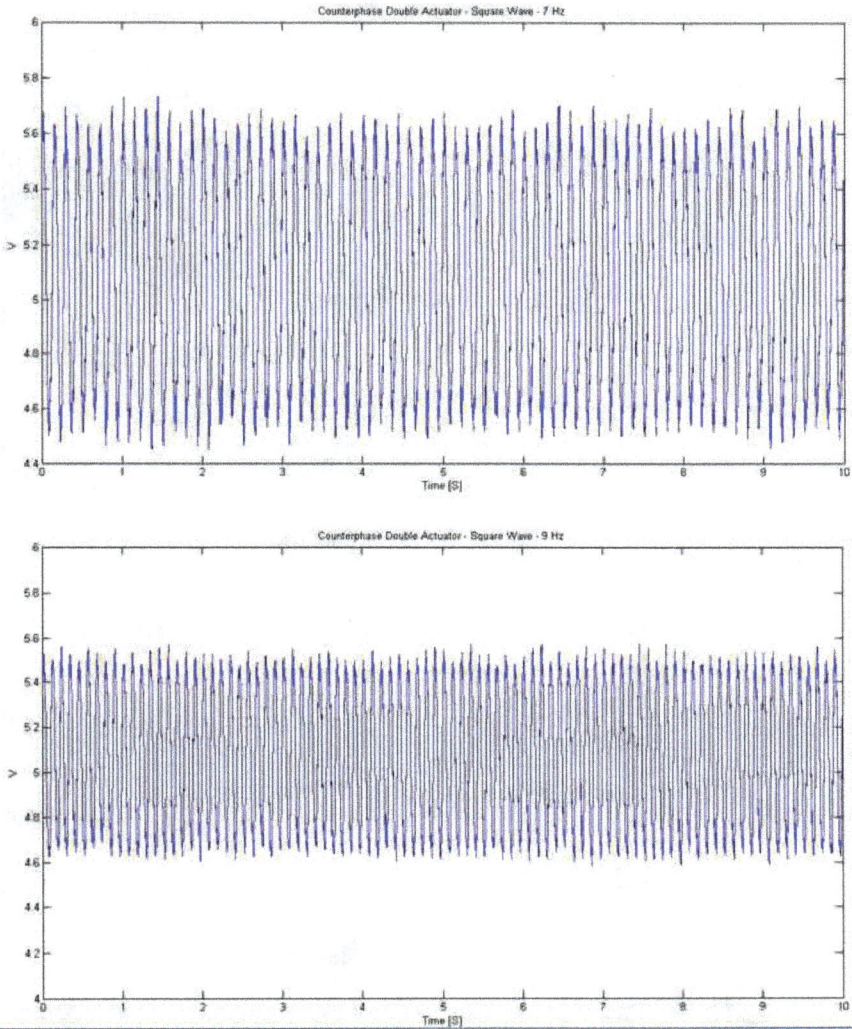

Fig. 5.37 Displacement of the internal pier of the circular structure excited by a pair of anti-phase electromechanical actuator driven by a sinusoidal waveform with different frequencies: (a) 3Hz, (b) 5Hz, (c) 7Hz, (d) 9Hz, (e) 11Hz, (f) 13Hz.

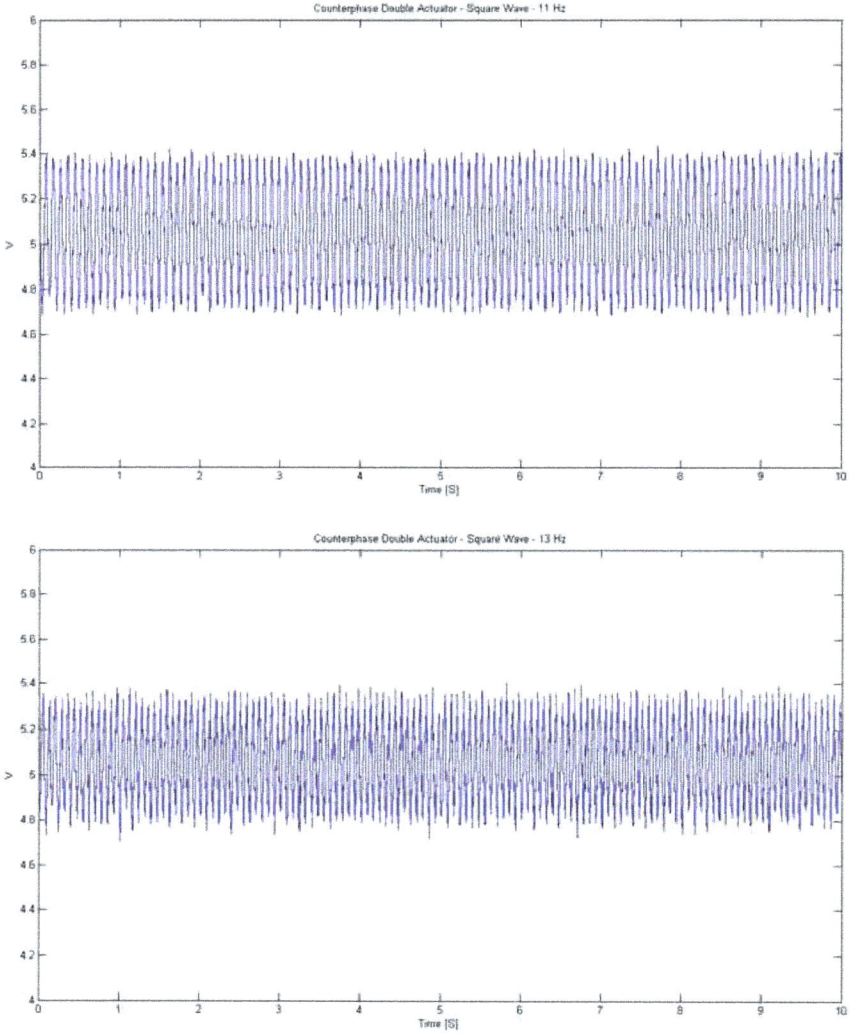

Fig. 5.38   Displacement of the internal pier of the circular structure excited by a pair of anti-phase electromechanical actuators driven by a sinusoidal waveform with different frequencies: (a) 3Hz, (b) 5Hz, (c) 7Hz, (d) 9Hz, (e) 11Hz, (f) 13Hz.

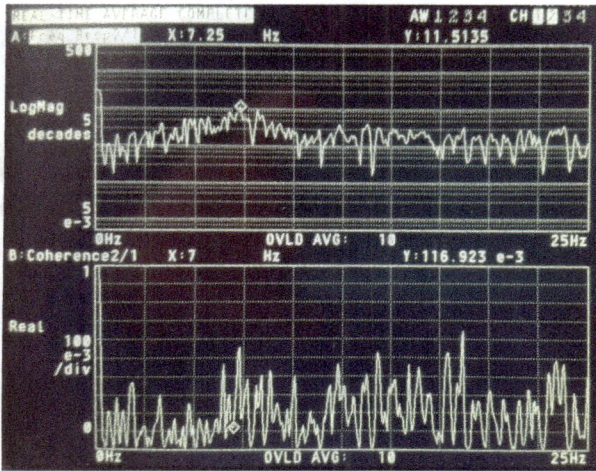

Fig. 5.39 Experimental estimation of the frequency response of the circular structure with respect to induced vibrations. Displacement is measured by means of the laser sensor located on one supporting pier. Upper panel: frequency response considering as input a sinusoidal waveform used to drive the actuation system and as output the displacement of the rectangular structure measured by the laser sensor. Lower panel: spectral coherence $C_{xy}(f)$.

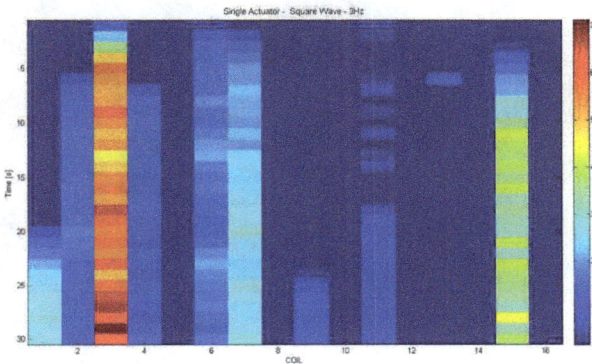

Fig. 5.40 Angular speed (Hz) of the sixteen coils - Excitation with a single actuator.

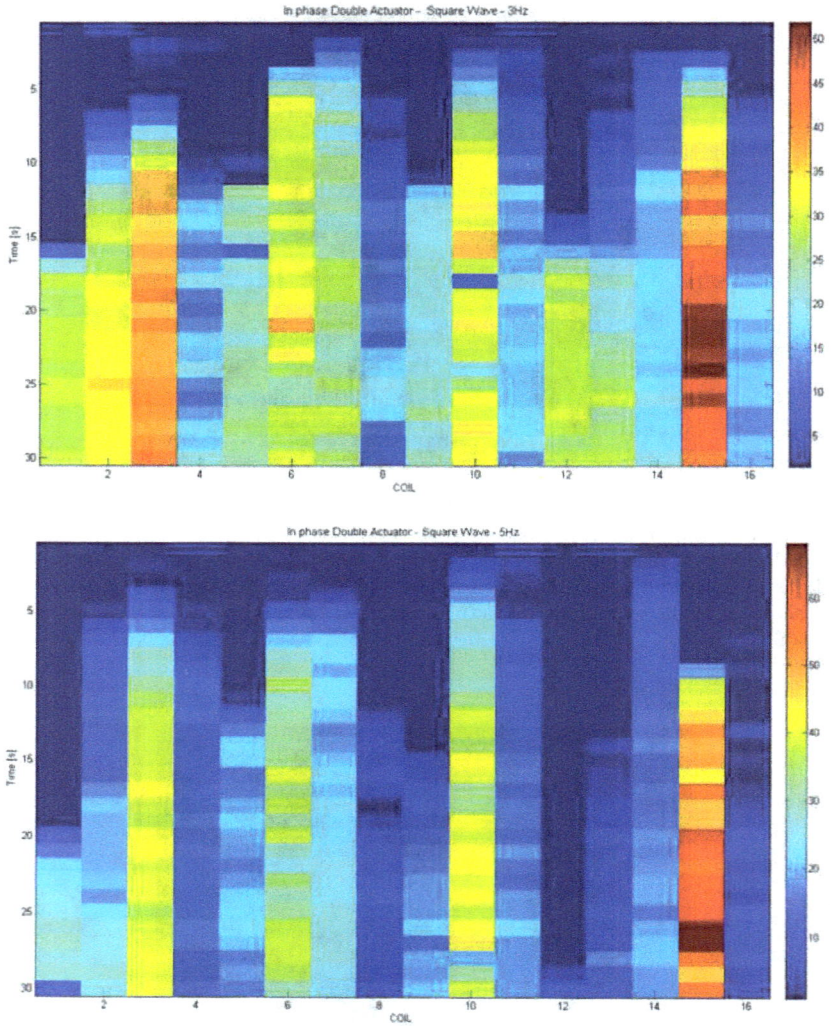

Fig. 5.41   Angular speed (Hz) of the sixteen coils - Excitation with two in-phase actuators driven by a square waveform with frequency: (a) $f = 3$Hz, (b) $f = 5$Hz.

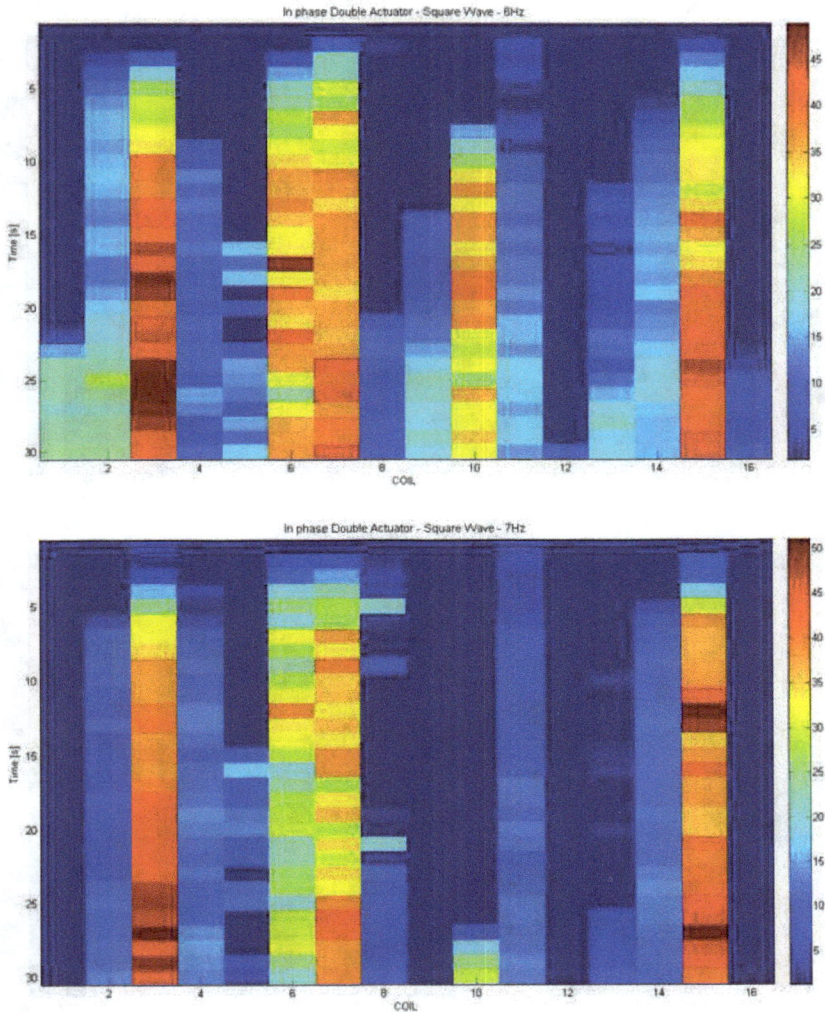

Fig. 5.42 Angular speed (Hz) of the sixteen coils - Excitation with two in-phase actuators driven by a square waveform with frequency: (a) $f = 6$Hz, (b) $f = 7$Hz.

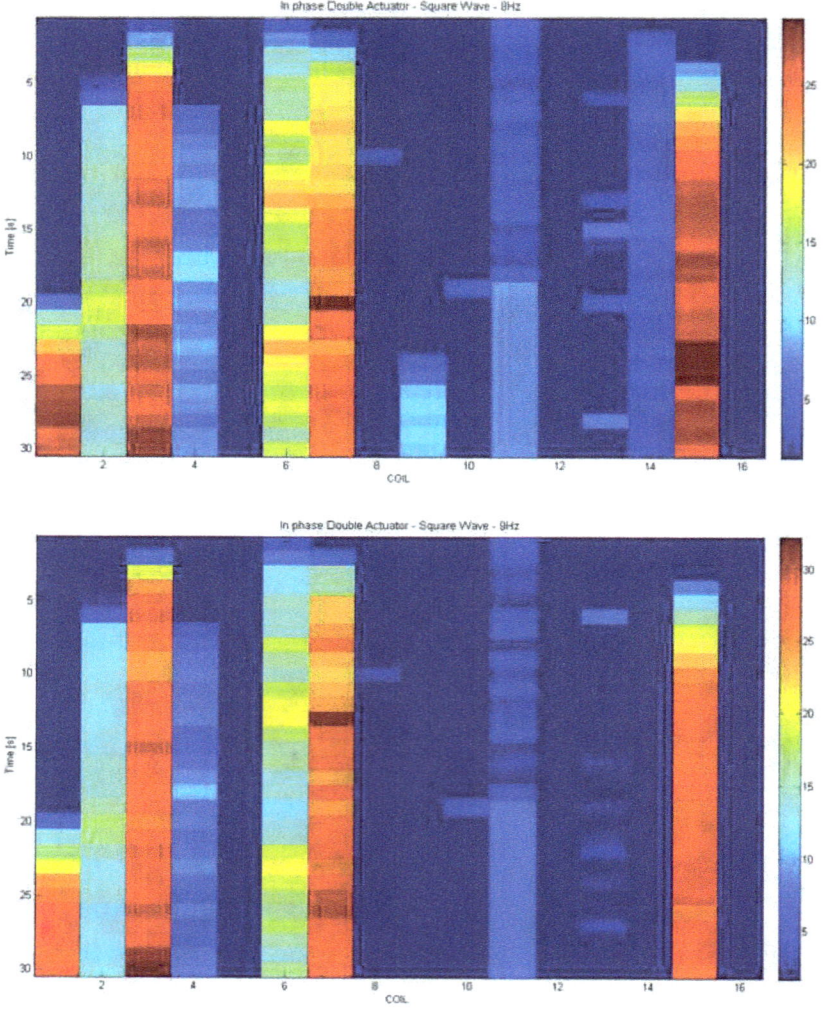

Fig. 5.43   Angular speed (Hz) of the sixteen coils - Excitation with two in-phase actuators driven by a square waveform with frequency: (a) $f = 8$Hz, (b) $f = 9$Hz.

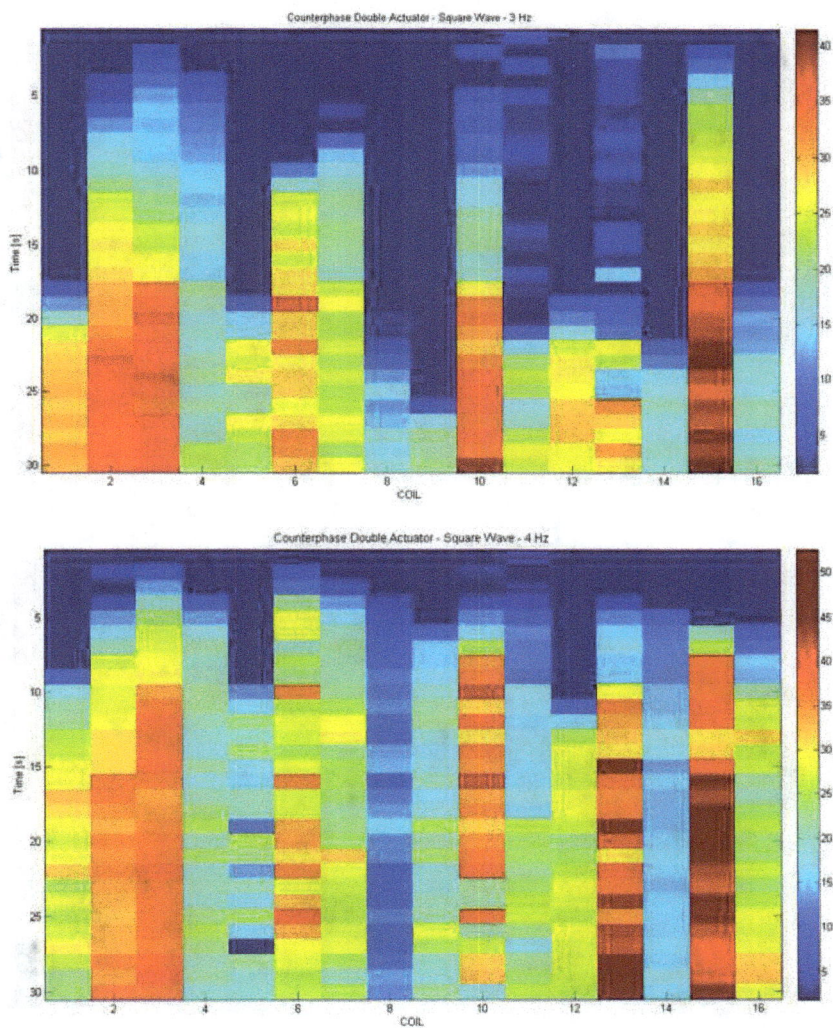

Fig. 5.44 Angular speed (Hz) of the sixteen coils - Excitation with two anti-phase actuators driven by a square waveform with frequency: (a) $f = 3$Hz, (b) $f = 4$Hz.

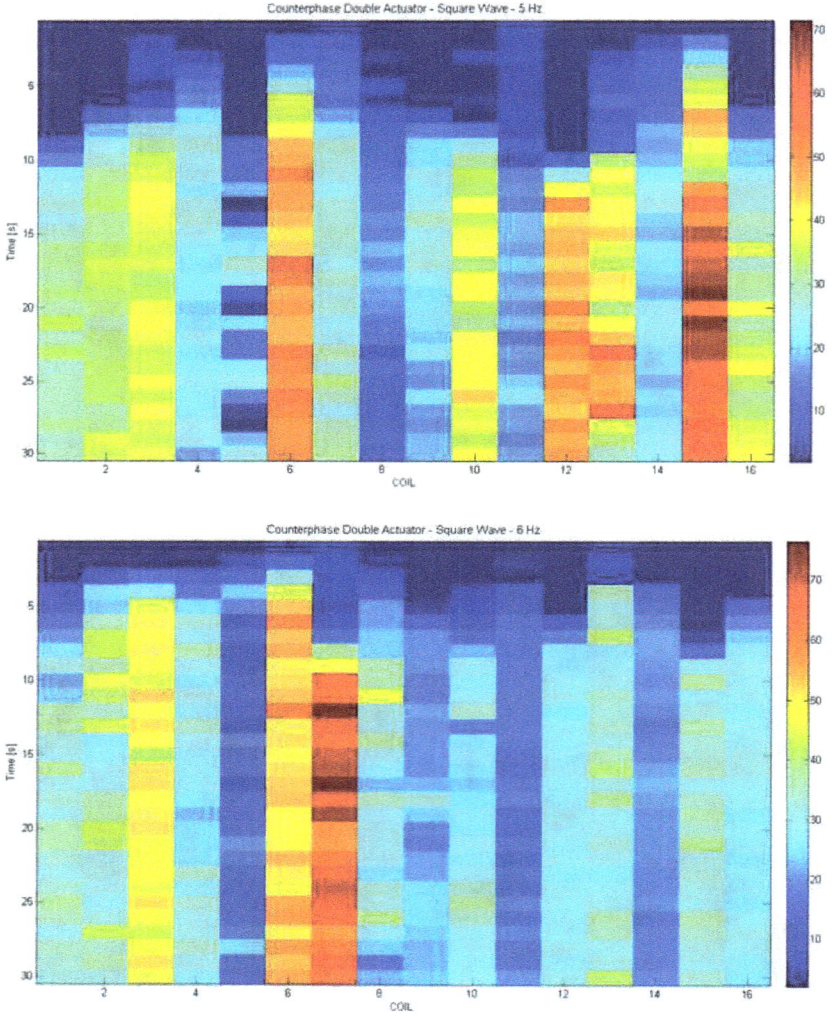

Fig. 5.45   Angular speed (Hz) of the sixteen coils - Excitation with two anti-phase actuators driven by a square waveform with frequency: (a) $f = 5$Hz, (b) $f = 6$Hz.

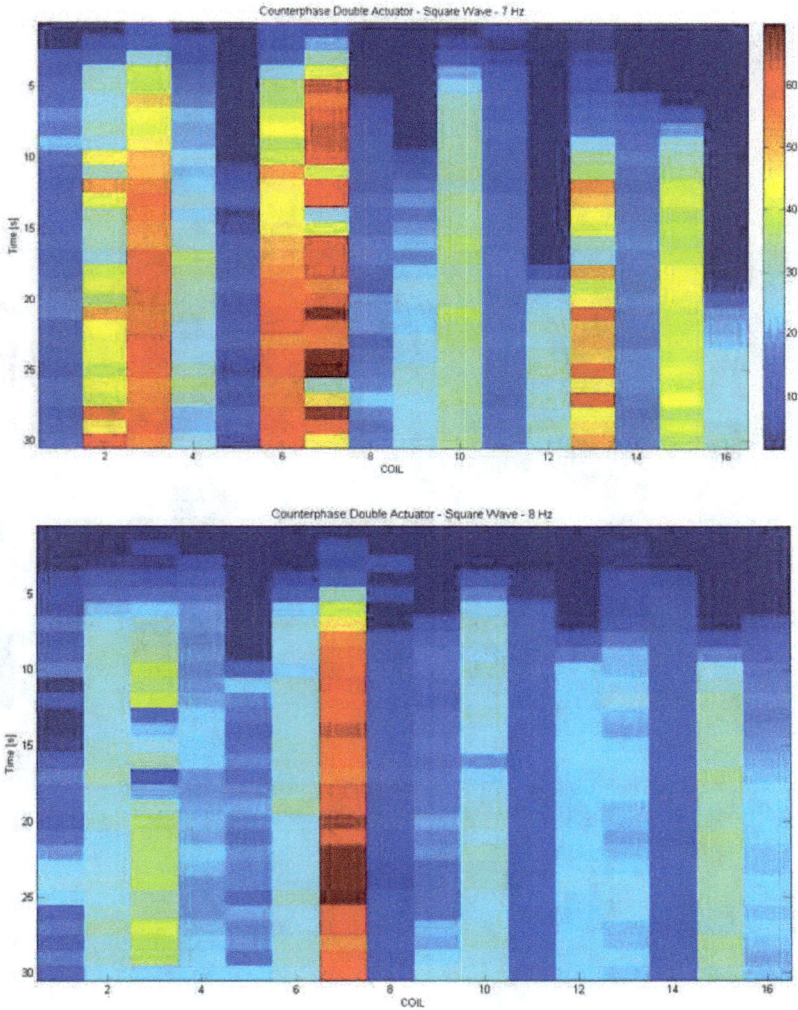

Fig. 5.46 Angular speed (Hz) of the sixteen coils - Excitation with two anti-phase actuators driven by a square waveform with frequency: (a) $f = 7$Hz, (b) $f = 8$Hz.

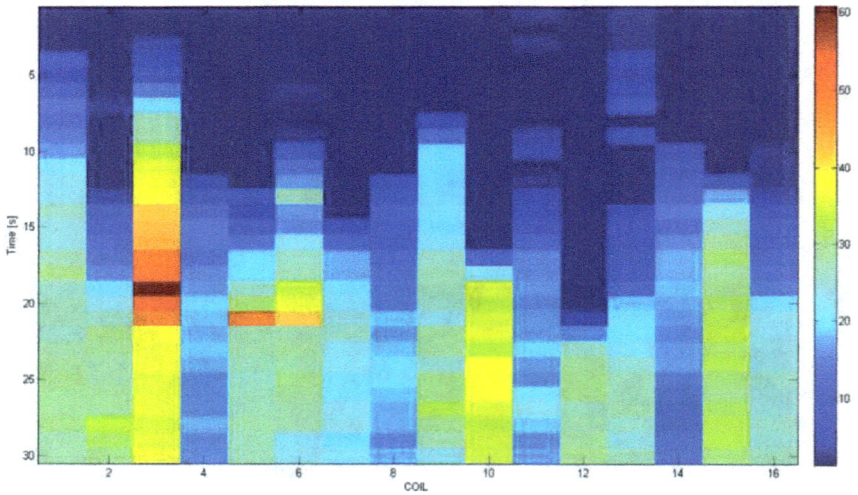

Fig. 5.47   Angular speeds (Hz) of the sixteen coils during a complete experiment. All the coils are rotating for $t > 21$s. From the color code, uniformity in speed can be deduced.

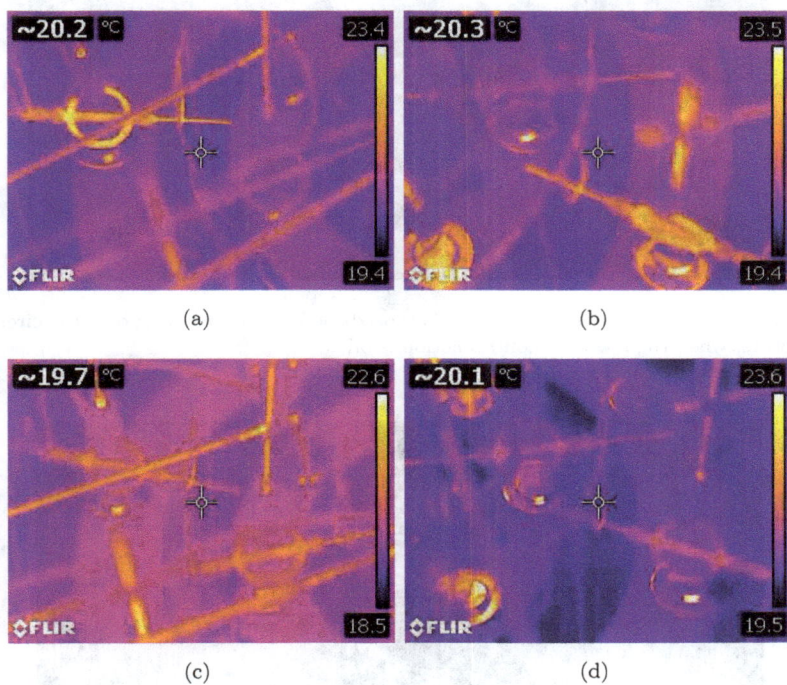

(a)   (b)

(c)   (d)

Fig. 5.48   Thermographic images for the circular structure: (a)-(b) regularizing control action is off, (c)-(d) regularizing control action is on.

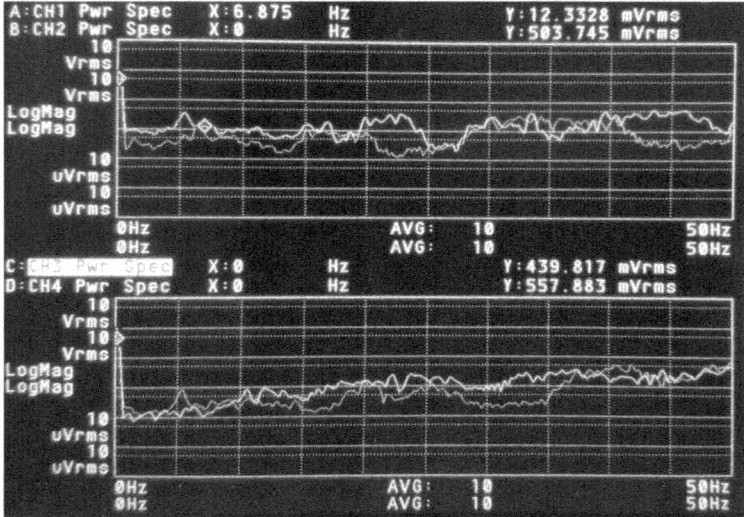

Fig. 5.49   Spectral analysis of the angular speeds of four coils rotating over the circular structure, when the low level control action is off.

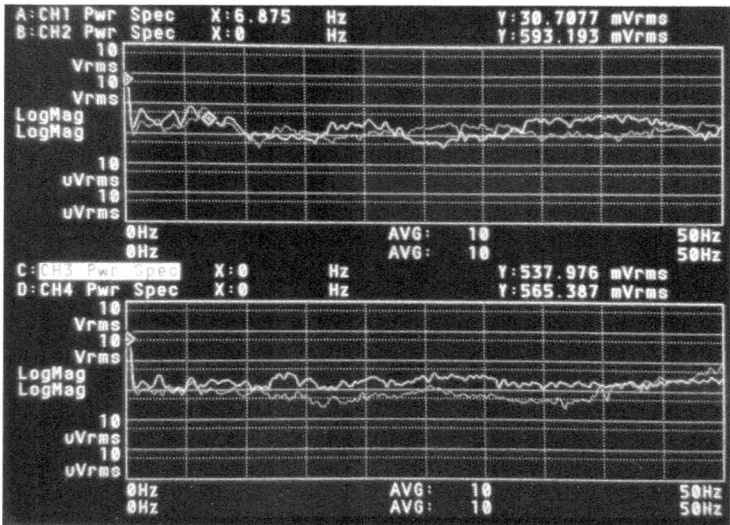

Fig. 5.50   Spectral analysis of the angular speeds of four coils rotating over the circular structure, when the low level control action is on.

Chapter 6

Active Imperfect Systems to Realize Chaotic Mechanical Oscillators

This chapter is devoted to describe two applications of the imperfect uncertain electromechanical systems investigated in this book. In particular, we are interested in showing that, despite the presence of imperfections, or better, thanks to them, complex behavior, ranging from limit cycle oscillations to chaotic vibrations can be obtained, even in small-scale systems.

6.1 A New Electromechanical Oscillator

In this section, a new electro-mechanical chaotic oscillator is presented. The system is based on the motion of the metal tip of a beam in a double-well potential generated by two magnets, and works thanks to the vibrations generated in the flexible mechanical structure by two rotating coils that produce noise-like signals. As the source of vibration is internal, the system may be considered an autonomous oscillator. Chaotic motion is experimentally observed and verified with a mathematical model of the phenomenon.

The nonlinear part of the system is based on a mechanism similar to that characterizing the Duffing oscillator [Ueda (1979); Erturk and Inman (2011)], which belongs to the family of those systems where the nonlinear behavior derives from a double-well potential, that is, the nonlinearity is such that two equilibria separated by a potential barrier coexist. Such systems are conceived as periodically forced mechanical structures composed of a flexible beam with a metal end, which is deflected towards two positions, generating a bistable configuration, induced by the presence of two magnets [Moon (2004); Owens et al. (2013)]. Usually, the complex behavior of such a mechanical system is induced by an external forcing signal, which is modeled as a sinusoidal signal with amplitude A and frequency f. Both A and f act as bifurcation parameters, whose variations lead to different nonlinear behaviors, widely studied in the literature [Sprott (2003); Ott (2002); Holmes (1979)], ranging from equilibrium to nonlinear periodical oscillations and chaos.

Unlike the classical magneto-elastic buckled beam [Moon (2004)], the scheme proposed here is based on an internally generated mechanical vibration, which is created by two coils rotating at different speeds. The effect of these rotating coils on the mechanical structure is to transfer both a vertical and horizontal oscillation to the flexible beam that is subjected to the interaction with two magnets. This is substantially different from the classical magneto-elastic buckled beam, where a shaker is needed to provide a sinusoidal forcing signal only in the horizontal direction [Moon (2004)]. The peculiarity of the system is, then, to exploit a noise-like mechanical, internally generated signal, a further confirmation of the role of imperfections in the generation of regular chaotic oscillations [Harada *et al.* (1996); Chotorlishvili *et al.* (2011); Siewe Siewe *et al.* (2010); Steur (2007); Bonato Altran *et al.* (2011)].

6.1.1 *The electro-mechanical setup*

The electro-mechanical system is shown in Fig. 6.1. The mechanical structure consists of one elastic trail, which is suspended between two fixed supports and hosts two coils and a portal where a flexible beam is hung. At the tip of the flexible beam a small metal plate is mounted, so that the beam is deflected within a magnetic field generated by two small neodymium magnets, each providing a magnetic force of 1.5kg, located on the bottom part of the system. Each coil, powered through the copper wires of the trail, constitutes the rotor of a simple DC motor with the stator realized through a neodymium magnet providing a magnetic force of 14kg, also located on the bottom part of the system. The rotation of the two coils is the source of a mechanical vibration which is fundamental to induce a chaotic motion to the beam tip. The schematic representation reported in Fig. 6.1 illustrates the sizes of the experimental setup and the location of the two pairs of magnets. Particulars of the structure are shown in Fig. 6.2.

The working principle of the system may be described by viewing it as formed by two interacting mechanical parts. The first is the flexible beam whose tip moves in a double well potential generated by the two small magnets. In the absence of vibrations, the tip goes to one of the two equilibrium positions of the field. The second mechanical part consists of the two coils and their support. During its rotation each coil is subjected to a friction which lifts it up along the slopes of the arcs where its pins are placed. When it suddenly goes down, a shock is transmitted to the supporting structure. As a consequence of this irregular motion, therefore, the rotation of the two coils vibrates the elastic trail. It is worth noticing that this phenomenon, which induces a sequence of beats on the structure, occurs at different times between the two coils. This is mainly due to their different rotation speeds, a consequence of the intrinsic imperfections.

The shocks on the mechanical vibrating trail, produced by the coils during their rotation, generates a two-component force which is transmitted to the portal and the flexible beam, making the motion of the tip more complex as its distance from the two magnets is now also a function of the portal vibration. The component of the vibration force along the vertical direction moves the portal up and down and, in so doing, modulates the magnetic effect on the tip of the flexible beam and changes the potential barrier to move from one equilibrium position to the other. When the potential barrier is decreased, the presence of the horizontal component acting at the same time induces the switch of the tip between the two stable positions.

6.1.2 *Mathematical model of the oscillator*

Our system hosts two coils, which, when rotating, cause the mechanical vibration of the whole structure. This latter can be viewed as a constrained beam, whose dynamical response to the input (the vibrations of the coils) also depends on the mechanical properties of the structure itself. As it will be shown in the following, a model only considering the dominant frequency of the oscillations in the horizontal and vertical directions is capable to capture the dynamics of the whole system, at the same time keeping the approach simple.

Following the approach presented in [Moon (2004)], we consider the motion of the tip of the flexible beam hanging on the trails and describe it by focusing on a dimensionless modal amplitude variable, x, and its derivative y. The tip motion is, therefore, described by the following equations:

$$\dot{x} = \kappa y$$
$$\dot{y} = \kappa(-\gamma y + \tfrac{1}{2}x(1 - x^2)F(\Delta v) - G(\Delta h)) \tag{6.1}$$

where γ is the damping factor, and κ is a temporal rescaling factor introduced to match the dimensionless model with the physical system. $F(\Delta v)$ and $G(\Delta h)$ are functions of the vertical and horizontal displacements of the support where the beam is hung and represent the influence on the

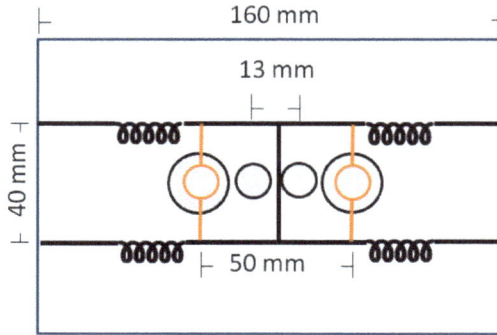

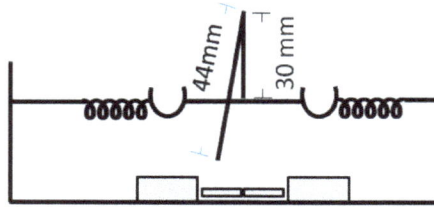

(a)

(b)

Fig. 6.1 The experimental setup: (a) schematic representation of the electro-mechanical system; (b) a picture of the setup.

tip motion of the vibrations induced by the rotation of the coils through the supporting structure. As mentioned, we consider the main (sinusoidal) components for the vertical and horizontal displacements, that is, $\Delta v \simeq A_v \sin(2\pi f_v t)$ and $\Delta h \simeq A_h \sin(2\pi f_h t)$, and, so, consider $F(\Delta v)$

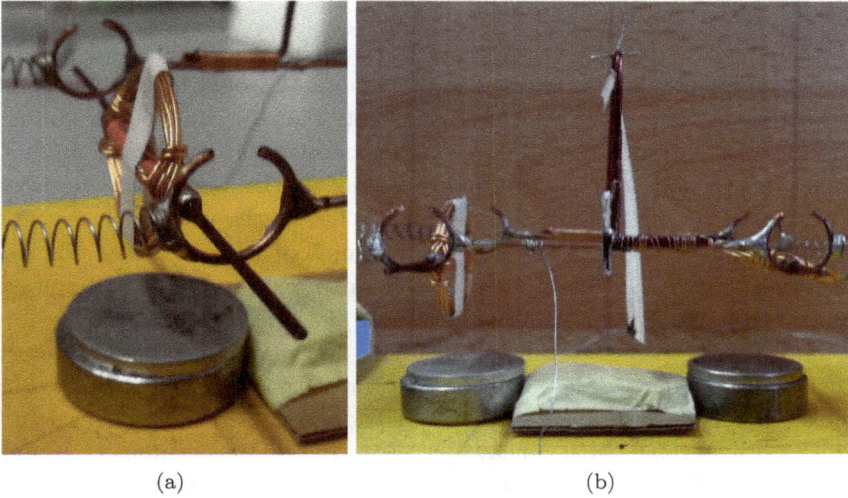

Fig. 6.2   Particulars of the mechanical structure: (a) support of one coil; (b) front view of the oscillating beam.

and $G(\Delta h)$ as functions of amplitude and frequency of the dominant component, that is, $F(\Delta v) \simeq F(A_v, f_v)$ and $G(\Delta h) \simeq G(A_h, f_h)$. We explicit these functions as

$$F(A_v, f_v) = A_v \sin 2\pi f_v t + 1$$

and

$$G(A_h, f_h) = A_h (2\pi f_h)^2 \cos 2\pi f_h t.$$

The mechanical vibrations induced by the pair of rotating coils, in fact, are encompassed in the mathematical model through $F(A_v, f_v)$ and $G(A_h, f_h)$. While the effect of the horizontal displacement Δh can be considered as that induced by an external forcing, like in the magneto-elastic buckled beam [Moon (2004)], the effect of the vertical displacement Δv is to modulate the strength of the magnetic force acting on the tip. In fact, even the vertical displacement, making the beam moving in a direction orthogonal to x, changes the tip position within the magnetic field induced by the two magnets. The effect is that, when the structure shifts upward, the beam tip departs from the magnets and their effect is mitigated, while the opposite holds when the structure moves downward.

6.1.3   *Experimental results*

The experimental setup, realized in our laboratory, only needs a constant
DC source to operate. It is interesting to note that the value of this voltage
supply is also an important bifurcation parameter for the system. In fact,
it controls the speed of the rotating coils and, so, the amount of vibrations
transmitted to the flexible beam. Chaotic motion is observed in a fairly
large range of this parameter. A first example of the chaotic motion observed is reported in Fig. 6.3, which refers to a value of the coil voltage
supply equal to $V_s = 420$mV. The trend of the position of the flexible
beam tip with respect to the rest position, $s(t)$, measured with a Baumer
OADM 12U6460/S35A laser distance sensor, is shown in Fig. 6.4(a), while
the attractor, reconstructed through embedding [Takens (1981); Abarbanel
(2012); Cook (1994)] with a delay $\tau = 100$ms from the measurement of $s(t)$,
is illustrated in Fig. 6.4(b). For the data in Fig. 6.3 a largest Lyapunov
exponent equal to $\Lambda_{max} = 0.0325$, confirming the chaotic nature of the
tip motion, has been obtained by the Rosenstein method [Rosenstein *et al.*
(1993)].

By varying the value of the voltage, the system operates in other re-
gimes: in particular, if V_s is increased, in the range of $V_s \in [480, 840]$mV
chaotic motion with the system switching from one scroll to the other is
observed. As an example, the chaotic motion reported in Fig. 6.4 refers to
the coil voltage supply equal to $V_s = 550$mV. The trend of the position $s(t)$
is shown in Fig. 6.4(a), while the attractor, reconstructed through embedding with a delay $\tau = 100$ms is illustrated in Fig. 6.4(b). For the data in
Fig. 6.4 a largest Lyapunov exponent is equal to $\Lambda_{max} = 0.1092$, once again
confirming that the chaotic nature of the tip motion has been obtained with
the Rosenstein method [Rosenstein *et al.* (1993)]. Furthermore, for values
close to the transition to this latter regime, intermittent behavior between
the two scrolls, with one visited more often than the other, is observed.
Finally, when $V_s > 840$mV the power provided to the two coils makes them
capable of overcoming friction and, as a consequence, they start to rotate
without hitting the trails and then without providing the two-components
vibration to the oscillating beam.

The system behavior has been characterized through a bifurcation diagram, where the local maxima of $s(t)$ are plotted with respect to V_s. The
diagram is shown in Fig. 6.5 and illustrates the transition from single to
double scroll at $V_s \approx 480$mV.

The experimental setup is also equipped with two other Baumer OADM

12U6460/S35A laser distance sensors used to measure the vibration of the structure by monitoring its vertical and horizontal displacements Δh and Δv. An example of the displacements produced in the mechanical structure obtained for $V_s = 550$mV is reported in Fig. 6.6. The vertical displacement Δv clearly shows higher amplitudes than the horizontal one. This observation is verified also by varying the coils supply voltage, and is at the basis of the emerging behavior of the system.

The measured signals have been used for the identification of the parameters of the model in Eqs. (6.1), and, in particular, the values of the frequency of vertical and horizontal displacements. From these measurements, performed at different values of the voltage supply, it is possible to observe that the frequency of the horizontal vibration ranges from 18Hz to 25Hz, while the frequency of the vertical vibration is confined within 8Hz to 10Hz. The values obtained for f_h and f_v have then been used along with the other parameters experimentally identified and constant with V_s ($\gamma = 0.33$, $A_h = 0.2$, $A_v = 1$ and $\kappa = \frac{1}{100}$) for the numerical integration of the model in Eqs. (6.1). Figure 6.7 show the reconstructed attractors from the numerical simulations with $V_s = 420$mV ($f_h = 18$Hz and $f_v = 8$Hz) and $V_s = 550$mV ($f_h = 24$Hz and $f_v = 10$Hz). The largest Lyapunov exponent has been calculated in both cases from the model as 0.0275 and 0.0779, respectively. The experimental and the numerical trends are very similar, thus showing the suitability of the proposed modeling.

6.2 A Driven Chaotic Rotating Coil

Let us consider now the simplest electromechanical configuration based on the coil-magnet system introduced in this book. The interest is to investigate the possibility to generate complex dynamics by using a single coil-magnet system.

In the experiment discussed in the previous section, a constant voltage power supply V_a has always been adopted. Now we consider the possibility of exciting the coils with a square wave signal, i.e. a signal providing the power supply which has a high and a low level. This means to excite the coil-magnet system with an external driving signal with a fixed amplitude and tunable frequency defined as

$$v(t) = \frac{V_a}{2} \left[(\text{sgn}\,(\sin(2\pi ft) + 1)) \right]$$

where f is the frequency and V_a is the amplitude.

Therefore the dynamics of the coil does depend on the frequency f which can be considered as a bifurcation parameter, and the complex behavior of the system can be characterized by means of the corresponding bifurcation diagram.

6.2.1 Driving a single coil-magnet system

In order to provide a switching power supply to the coil-magnet system, we need to design a suitable circuit to provide the needed current to the coil. We focused on the scheme reported in Fig. 6.8 in which the power transistor $2N3055$ is used as a switch. This component is able to provide a power up to 115W with collector currents in the order of 15A.

For our purpose we will use a power supply providing a constant voltage of 2V to polarize the transistor, and a function generator to realize the square wave. In this way, the coil will be powered by the power supply only when the square wave generated by the function generator is at the high level. In this case, in fact, the transistor is turned on and the power goes to the coil-magnet system. Otherwise, when the square wave is at the low level, no current flows through the transistor and, therefore, the coil-magnet system is not powered. In order to change the frequency, it is sufficient to act on the settings of the function generator which drives the switch.

6.2.2 Mathematical model of the driven chaotic rotating coil and experimental results

From a mathematical point of view, the introduction of a square wave signal to power the coil-magnet system can be easily incorporated in the model. Let us consider again Eqs. (3.1) modeling the single coil-magnet system. The current I is redefined as

$$I = \frac{V_a(t) - YSB\sin(X)}{R(X)}$$

where the term

$$V_a(t) = \frac{V_a}{2}\left[(\mathrm{sgn}\left(\sin(2\pi ft) + 1\right))\right]$$

is included.

The numerical bifurcation diagram obtained by varying f is reported in Fig. 6.9. It has been obtained by calculating the frequency $f_c = \frac{Y}{2\pi}$ of the coil rotation and retrieving the corresponding local maxima. As it can be observed for values of f below 0.5Hz the local maxima are spread over a large interval of values, thus indicating the presence of a chaotic rotation of the coil over its support for a significant range of f. Moreover, increasing f leads to a condition in which the coils are constantly powered and, in fact, the local maxima of frequency f_c settle in a narrower range, thus indicating the transition towards a periodic motion.

These numerical results make evident the possibility to realize a further mechanical chaotic oscillator by using only the dynamics of a single coil, with a suitable excitation.

The experimental setup is reported in Fig. 6.10 and it is equipped with the optical sensors described in Chapter 3 to evaluate the angular speed of the coil. The frequency of the power supply is varied in the same range as used for the numerical analysis. The experimental bifurcation diagram obtained from the datasets acquired by calculating the local maxima of the temporal evolution of the angular speed, retrieved as discussed in Chapter 5, is shown in Fig. 6.11.

A qualitative agreement between the numerical results and the experimental trend can be appreciated. In fact, in the same range of the frequency f, the same behavior can be observed both in Figs. 6.9 and 6.11. This last result emphasizes the possibility to generate experimentally and in a very simple way an electromechanical oscillator with complex behavior that can be controlled by the power supply frequency.

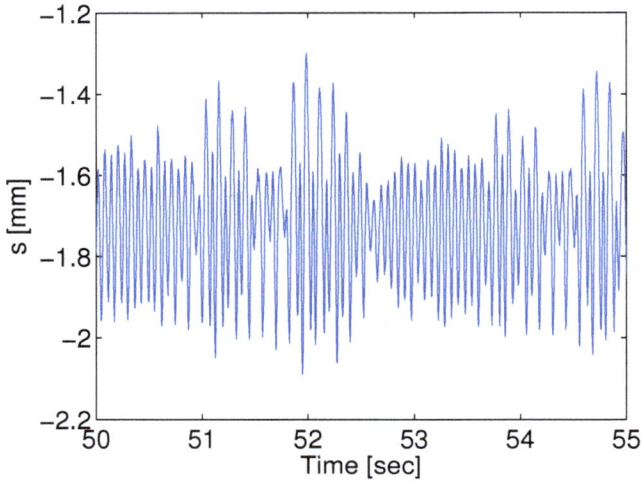

(a)

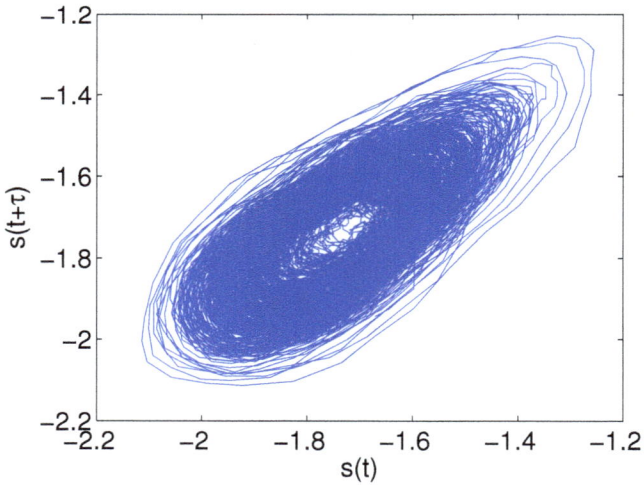

(b)

Fig. 6.3 Experimental results with $V_s = 420$mV: (a) displacement of the beam tip $s(t)$ with respect to the rest position, and (b) reconstructed attractor for a time delay $\tau = 100$ms from the acquired displacement of the beam tip.

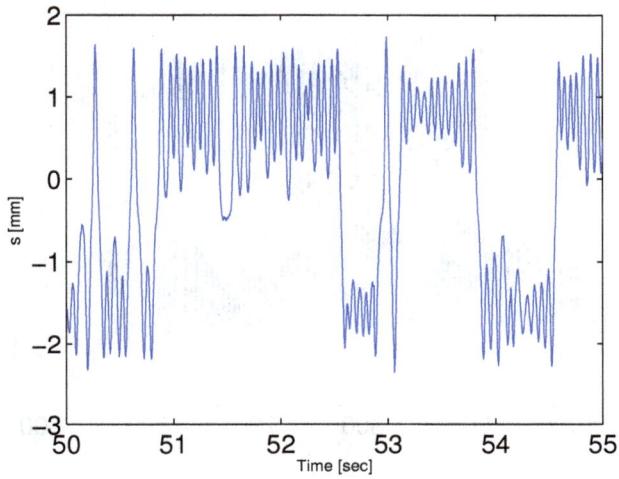

(a)

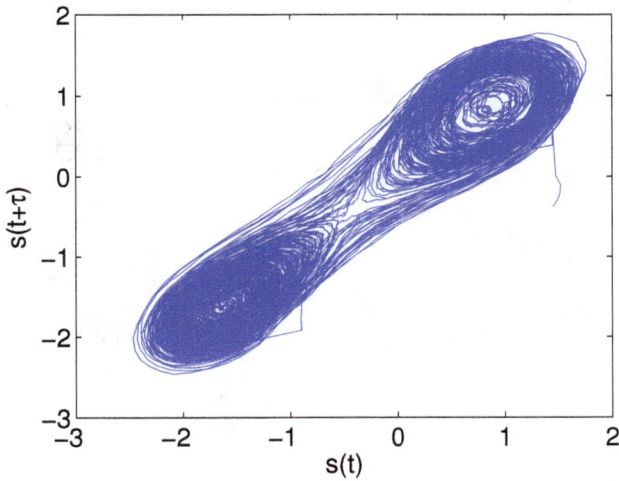

(b)

Fig. 6.4 Experimental results with $V_s = 550$mV: (a) displacement of the beam tip $s(t)$ with respect to the rest position, and (b) reconstructed attractor for a time delay $\tau = 100$ms from the acquired displacement of the beam tip.

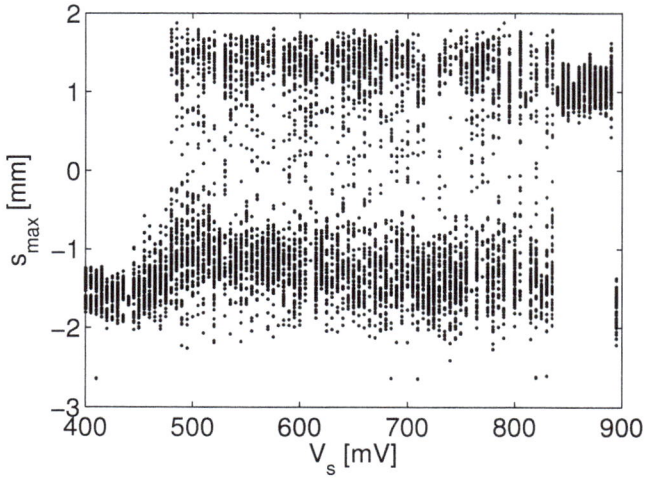

Fig. 6.5　Experimental bifurcation diagram of the tip motion. Local maxima of $s(t)$ are plotted with respect to V_s.

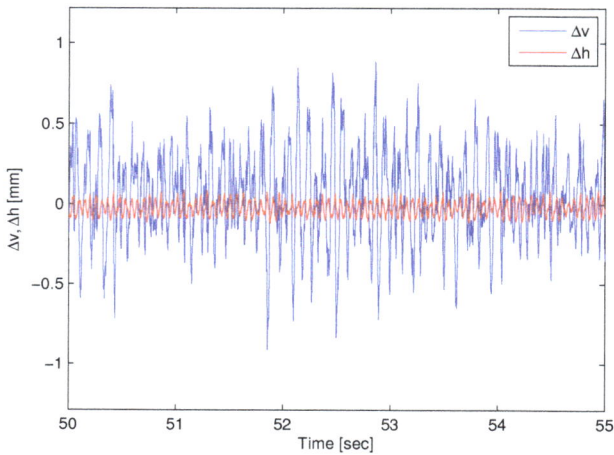

Fig. 6.6　Vertical and horizontal displacements of the oscillating structure with respect to the rest position at $V_s = 1.5V$.

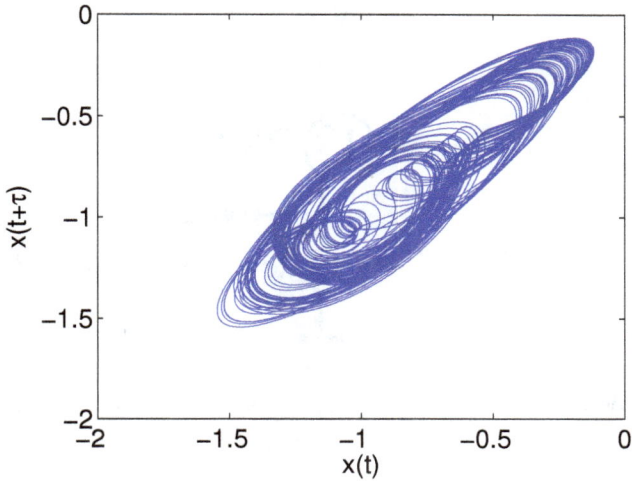

(a)

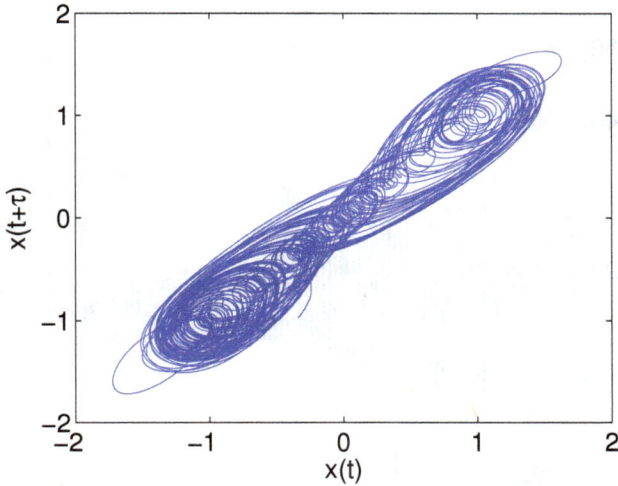

(b)

Fig. 6.7   Reconstructed attractors for a time delay $\tau = 100$ms from numerical integration of the model in Eqs. (6.1): (a) $V_s = 420$mV, and (b) $V_s = 550$mV.

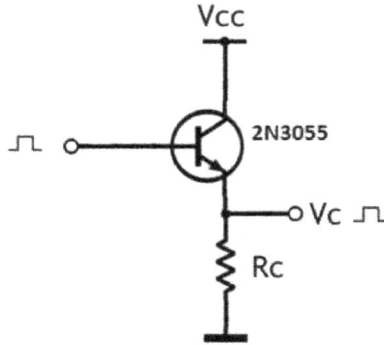

Fig. 6.8   Electrical scheme of the power supply based on the 2*N*3055 transistor connected as a switch.

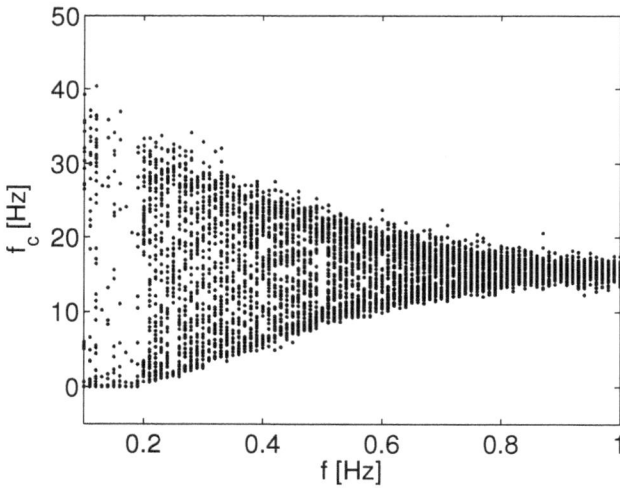

Fig. 6.9   Numerical bifurcation diagram with respect to the frequency f. Local maxima of $f_c = \frac{Y}{2\pi}$ are plotted for each value of the bifurcation parameter.

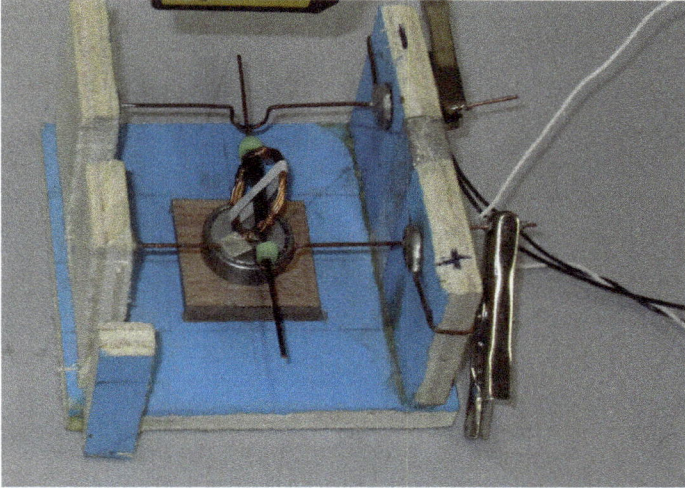

Fig. 6.10   Experimental setup consisting of a single coil-magnet system powered by a time-varying voltage.

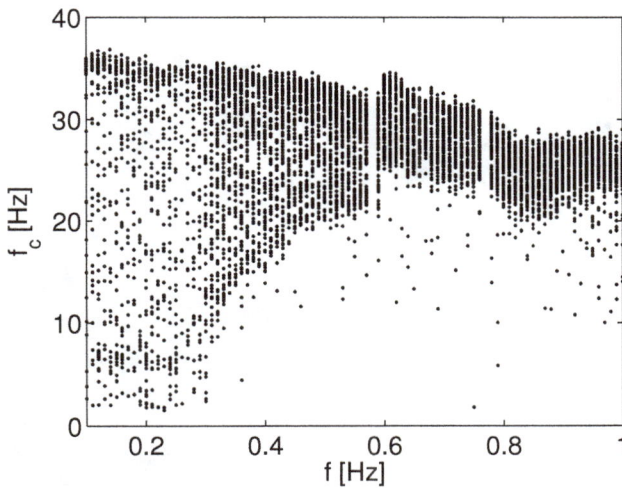

Fig. 6.11   Experimental bifurcation diagram with respect to the frequency f. Local maxima of the angular speed detected by the optical sensor are plotted for each value of the bifurcation parameter.

Chapter 7

Conclusions

The main aim of this monograph has been to emphasize the role of a class of nonlinear dynamical systems that have been named imperfect uncertain systems. Indeed, this class of systems has intrinsic capabilities of self-organization related to the high degree of nonlinearity included. Moreover, the self-organizing properties of this class of systems are derived from the interactions that continuously convey the information inside them.

The systems considered in this book are widely diffused. Examples of imperfect uncertain systems can be found in engineering structures and equipments and in many real-life dynamical systems. Moreover, such systems require a hybrid characterization, on the basis of both mathematical models and the empirical observations investigated by means of experimental data.

The possibility of controlling the behavior of such systems has been also investigated. The use of control strategies based on closed-loop vibrational control signals has been discussed, showing their effectiveness in allowing the correct and regular operativeness of the system. In order to remark both how to get models and how to establish the most suitable control strategy, two particular electromechanical structures have been studied in detail and their peculiarities have been widely discussed, also with reference to their geometrical properties.

Larger scale systems can be also considered in the framework of our studies. Preliminary experimental results obtained in coupled structures, such as that reported in Fig. 7.1, clearly show that the increasing number of units tends to improve the performance of the considered structures. It appears that systems in the class of nonlinear imperfect uncertain systems do work better if their complexity increases!

The previous concepts have been discussed step by step in the book.

Fig. 7.1   Two structures with five coils coupled through two springs.

In Chapter 6 the practical use of the previously considered systems has been discussed, taking into account the simplest configurations. In fact, by using a simple configuration of two coupled coil-magnet systems, or even a single coil-magnet system, a new class of mechanical oscillator has been introduced, showing the possibility to obtain complex behavior, like chaotic oscillations.

In summary, we have proposed a complete monograph dealing with the peculiarities of a new class of nonlinear dynamical systems, and have introduced the strategies for their modeling, and for favoring their self-organization. The fundamental role of the experiments to carry out our study has been emphasized. Indeed, what has been reported in this book is only what we consider the fundamental items to introduce the subject. Readers are encouraged to investigate and to discover the role of imperfect uncertain systems in their research fields. People working in industries can retrieve the fact that a lot of equipments do work thanks to uncertainty and imperfections.

Therefore, we hope that the interest in this subject can grow. Further efforts can be devoted to formalize in detail from a mathematical point of view what has been only preliminarily discussed.

Chapter 8

Epilogue

8.1 A Route to Engineering Studies

Let us consider the main subjects we have dealt with in this book. They should be considered as forming a comprehensive paradigm to study and control dynamical systems. In fact, taking into account the various items we have found and discussed, many concepts of electrical, electronic and control engineering are considered. Therefore, the efforts of our study have been devoted also to the educational perspective, so as to emphasize the importance of following the guidelines of this book for engineering studies [Awrejcewicz and Koruba (2012)].

Let us use the following route:

(1) The very simple working mechanism of electromechanical systems is emphasized from the behavior of the single coil. Each student can experiment by using only a coil, a magnet, an almost exhausted battery and a simple mechanical support (see Fig. 1.9). He can derive the model to describe the behavior of the coil and to understand the easy principles of the electromagnetic system. The principle of motion of the motor can be learnt.

(2) Making more than one coil in a single structure (see Fig. 1.8), the student can perceive the intuitive principle of systems cooperation.

(3) The mechanical vibration can be observed and the relationship between the power supply of the coils and the vibration can be found. Moreover the natural frequencies of a mechanical structure can be qualitatively observed.

(4) The use of a very simple equipment to measure the coil angular speeds (see Fig. 3.4) can be considered. This means to stimulate people to observe and measure. An oscilloscope or an application in a smartphone

can be used to observe low voltage signals such as those taken from a sensor. Therefore, the student can observe and perform measurements of coils speed.

(5) Moreover the complexity of the mechanical structure can be increased and therefore nonlinear interactions can be considered. The geometrical shape of the structure (see Fig. 1.8(b)) can be changed and experiments in order to identify which is the more flexible or the more rigid can be performed. This can be done also by means of laser sensors that provide a measure of the structure vibrations (see Fig. 5.3).

(6) When we have a system, a model must be used to control it. We can have mathematical or empirical models. The complex mechanical structures induce people to understand the vibration principles.

(7) Until now the performed operations have been done with one only external signal: the power voltage! Then the student can use external output mechanical signals to force the structure. He can use an external electromagnetic actuator (see Fig. 5.8). Moreover, to give some power to it, an easy power amplifier can be used. Therefore, the student is introduced to electronics: to build a conditioning amplifier for the sensors, and to build a power amplifier for the actuators. The main principles of electronics can be introduced at this point. Furthermore, what happens if batteries are completely exhausted? It is necessary to learn how to build a power supply system!

(8) If the student has to stimulate the electromagnetic actuator he requires a sinusoidal wave: an oscillator must be designed and built. It can be realized by using specific integrated circuits (i.e. LM555) which allow a change of the frequency. At this point, a new circuit is coupled to the whole structure.

(9) By using various frequencies in order to excite the system, the main natural frequencies of the electromechanical structures can be discovered using the electromechanical sensors. Therefore, a complete study about the dynamics of the structure can be performed. The qualitative study can be made quantitative and the student can clearly understand what is happening.

(10) At this point, instead of using a fixed frequency stimulus, an oscillation that covers a wide range of frequencies can be realized. The student is introduced to the implementation of chaotic circuits and their use to stimulate the structures driving the electromechanical actuators with the aim of discovering directly the natural frequency of the system. The student will appreciate which are the natural

frequencies analyzing the output of the different sensors located on the structure.

(11) As we said in the book, the power supply of the coil must be kept low. This is often a quite general question in dynamical systems: how to save energy? Answering this question means to solve an optimization problem, therefore the student is stimulated to approach this new aspect.

(12) Taking the power supply of the platform as small as possible, we require to stabilize the system. We plan to ensure that all the coils are working and that the coils do work with regularity. The student must fix these concepts in order to enhance the system performance.

(13) At this point, the concept of feedback can be introduced. The student can see that if some coils are not working, we can sense their behavior and perform actions in order to guarantee the right behavior for all of them. This can be done by using the electromechanical actuators. Indeed at this point we have to propose a control algorithm. In particular we introduced the concept of closed-loop vibrational control, emphasizing the main purpose of the feedback by switching on or off the actuators signal. This can be implemented by using a simple microcontroller.

(14) The student can discover how the control action can be performed by means of low-cost systems, like Arduino or STM Nucleo.

(15) The sensory system checks if the platform does or does not work properly and the control system sends the control signals to the actuators in order to provide suitable broad spectrum mechanical oscillations.

Therefore, summarizing the essential items provided by this book, we have:

- to understand principles of electromechanical systems, such as the coils and the electromechanical actuators;
- to apply simple sensors, like the optical sensor for monitoring angular speeds, or laser sensors to measure vibrations;
- to understand principles of mechanical dynamics;
- to design electrical circuits, such as power supply, power amplifiers, conditioning circuits and chaotic circuits;
- to formulate dynamical models and perform numerical simulations;
- to control complex systems;
- to build models and realize experiments in order to understand the role of imperfect uncertain systems.

8.2   Towards an Electromechanical Spiking Neuron

Let us conclude by considering again the mathematical model of the coil given in Eqs. (3.1). The numerical simulation of this system produces the trends of the current shown in Fig. 8.1(a). The experimental result referring to the same case showed a good agreement, as reported in Fig. 8.1(b).

Consider now the Hindmarsh-Rose (HR) model of a neuron described by the following set of ordinary differential equations:

$$\dot{x} = y - ax^3 + b^2 - z + I$$
$$\dot{y} = c - dx^2 - y \dot{z} = r \left[s(x - \lambda) - z \right] \tag{8.1}$$

Performing a numerical integration of system (8.1) fixing $a = 1$, $b = 3$, $c = 1$, $d = 5$, $I = 3.281$, $\lambda = -1.6$, $r = 0.002$, and $s = 4$, it is possible to retrieve the spiking behavior of the membrane potential in a neuron, as modeled by state variable $x(t)$, and reported in Fig. 8.1(c).

Now, let us compare the three figures, namely, the current flowing through the coil in numerical (Fig. 8.1(a)) and experimental (Fig. 8.1(b)) conditions and the membrane potential (Fig. 8.1(c)). The coil has a spiking bursting behavior like the HR neuron.

Can the dynamics of a coil lead us to the characterization of an electromechanical spiking neuron?

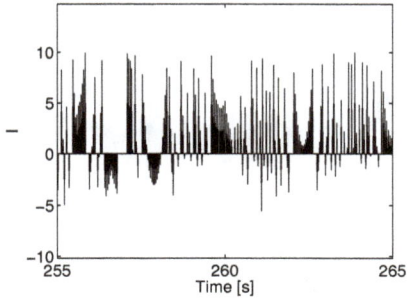

(a)

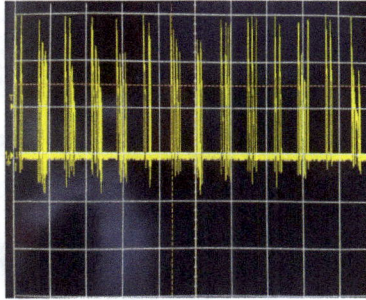

(b)

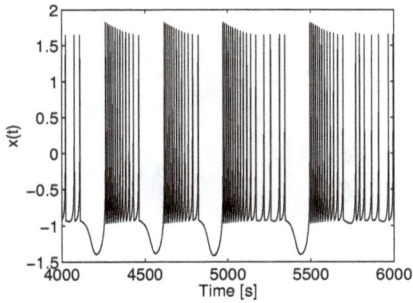

(c)

Fig. 8.1   Bursting behavior observed in (a) simulation of the current flowing in a single coil as modeled in Eqs. (3.1), (b) experimental observation of the current flowing in a single coil and (c) membrane potential of a neuron (variable $x(t)$ in Eqs. (8.1)).

Appendix A

Appendix

In this appendix we report the code of the microcontroller implementing the switching on and off control actions.

A.1 Microcontroller Code

File `main.c`

```
#include "stm32f3xx_hal.h"
ADC_HandleTypeDef hadc1;

TIM_HandleTypeDef htim1;
TIM_HandleTypeDef htim2;

uint16_t data, cont=0, nPeak=0, flag=0, nP2=0;
void SystemClock_Config(void);
static void MX_GPIO_Init(void);
static void MX_ADC1_Init(void);
static void MX_TIM1_Init(void);
static void MX_TIM2_Init(void);

int main(void)
{

  HAL_Init();
  SystemClock_Config();
  MX_GPIO_Init();
  MX_ADC1_Init();
  MX_TIM1_Init();
  MX_TIM2_Init();
  HAL_TIM_Base_Start_IT(&htim1);
  HAL_TIM_PWM_Start(&htim2, TIM_CHANNEL_1);
  HAL_TIM_PWM_Start(&htim2, TIM_CHANNEL_2);
  while (1)
  {

  }
}

void SystemClock_Config(void)
{
```

```
  RCC_OscInitTypeDef RCC_OscInitStruct;
  RCC_ClkInitTypeDef RCC_ClkInitStruct;
  RCC_PeriphCLKInitTypeDef PeriphClkInit;

  RCC_OscInitStruct.OscillatorType = RCC_OSCILLATORTYPE_HSI;
  RCC_OscInitStruct.HSIState = RCC_HSI_ON;
  RCC_OscInitStruct.HSICalibrationValue = 16;
  RCC_OscInitStruct.PLL.PLLState = RCC_PLL_ON;
  RCC_OscInitStruct.PLL.PLLSource = RCC_PLLSOURCE_HSI;
  RCC_OscInitStruct.PLL.PLLMUL = RCC_PLL_MUL16;
  HAL_RCC_OscConfig(&RCC_OscInitStruct);

  RCC_ClkInitStruct.ClockType = RCC_CLOCKTYPE_SYSCLK|RCC_CLOCKTYPE_PCLK1;
  RCC_ClkInitStruct.SYSCLKSource = RCC_SYSCLKSOURCE_PLLCLK;
  RCC_ClkInitStruct.AHBCLKDivider = RCC_SYSCLK_DIV1;
  RCC_ClkInitStruct.APB1CLKDivider = RCC_HCLK_DIV2;
  RCC_ClkInitStruct.APB2CLKDivider = RCC_HCLK_DIV1;
  HAL_RCC_ClockConfig(&RCC_ClkInitStruct, FLASH_LATENCY_2);

  PeriphClkInit.PeriphClockSelection = RCC_PERIPHCLK_TIM1|RCC_PERIPHCLK_ADC12;
  PeriphClkInit.Adc12ClockSelection = RCC_ADC12PLLCLK_DIV1;
  PeriphClkInit.Tim1ClockSelection = RCC_TIM1CLK_HCLK;
  HAL_RCCEx_PeriphCLKConfig(&PeriphClkInit);

  __SYSCFG_CLK_ENABLE();

}

void MX_ADC1_Init(void)
{

  ADC_ChannelConfTypeDef sConfig;

  hadc1.Instance = ADC1;
  hadc1.Init.ClockPrescaler = ADC_CLOCK_ASYNC;
  hadc1.Init.Resolution = ADC_RESOLUTION12b;
  hadc1.Init.ScanConvMode = ADC_SCAN_DISABLE;
  hadc1.Init.ContinuousConvMode = DISABLE;
  hadc1.Init.DiscontinuousConvMode = DISABLE;
  hadc1.Init.ExternalTrigConvEdge = ADC_EXTERNALTRIGCONVEDGE_NONE;
  hadc1.Init.DataAlign = ADC_DATAALIGN_RIGHT;
  hadc1.Init.NbrOfConversion = 1;
  hadc1.Init.DMAContinuousRequests = DISABLE;
  hadc1.Init.EOCSelection = EOC_SINGLE_CONV;
  hadc1.Init.LowPowerAutoWait = DISABLE;
  hadc1.Init.Overrun = OVR_DATA_OVERWRITTEN;
  HAL_ADC_Init(&hadc1);

  sConfig.Channel = ADC_CHANNEL_1;
  sConfig.Rank = 1;
  sConfig.SingleDiff = ADC_SINGLE_ENDED;
  sConfig.SamplingTime = ADC_SAMPLETIME_19CYCLES_5;
  sConfig.OffsetNumber = ADC_OFFSET_NONE;
  sConfig.Offset = 0;
  HAL_ADC_ConfigChannel(&hadc1, &sConfig);

}

void MX_TIM1_Init(void)
{
```

```
    TIM_ClockConfigTypeDef sClockSourceConfig;
    TIM_MasterConfigTypeDef sMasterConfig;

    htim1.Instance = TIM1;
    htim1.Init.Prescaler = 63;
    htim1.Init.CounterMode = TIM_COUNTERMODE_UP;
    htim1.Init.Period = 999;
    htim1.Init.ClockDivision = TIM_CLOCKDIVISION_DIV1;
    htim1.Init.RepetitionCounter = 0;
    HAL_TIM_Base_Init(&htim1);

    sClockSourceConfig.ClockSource = TIM_CLOCKSOURCE_INTERNAL;
    HAL_TIM_ConfigClockSource(&htim1, &sClockSourceConfig);

    sMasterConfig.MasterOutputTrigger = TIM_TRGO_RESET;
    sMasterConfig.MasterOutputTrigger2 = TIM_TRGO2_RESET;
    sMasterConfig.MasterSlaveMode = TIM_MASTERSLAVEMODE_DISABLE;
    HAL_TIMEx_MasterConfigSynchronization(&htim1, &sMasterConfig);

}

void MX_TIM2_Init(void)
{

    TIM_ClockConfigTypeDef sClockSourceConfig;
    TIM_MasterConfigTypeDef sMasterConfig;
    TIM_OC_InitTypeDef sConfigOC;

    htim2.Instance = TIM2;
    htim2.Init.Prescaler = 15999;
    htim2.Init.CounterMode = TIM_COUNTERMODE_UP;
    htim2.Init.Period = 999;
    htim2.Init.ClockDivision = TIM_CLOCKDIVISION_DIV1;
    HAL_TIM_Base_Init(&htim2);

    sClockSourceConfig.ClockSource = TIM_CLOCKSOURCE_INTERNAL;
    HAL_TIM_ConfigClockSource(&htim2, &sClockSourceConfig);

    HAL_TIM_PWM_Init(&htim2);

    sMasterConfig.MasterOutputTrigger = TIM_TRGO_RESET;
    sMasterConfig.MasterSlaveMode = TIM_MASTERSLAVEMODE_DISABLE;
    HAL_TIMEx_MasterConfigSynchronization(&htim2, &sMasterConfig);

    sConfigOC.OCMode = TIM_OCMODE_PWM1;
    sConfigOC.Pulse = 0;
    sConfigOC.OCPolarity = TIM_OCPOLARITY_HIGH;
    sConfigOC.OCFastMode = TIM_OCFAST_DISABLE;
    HAL_TIM_PWM_ConfigChannel(&htim2, &sConfigOC, TIM_CHANNEL_1);

    HAL_TIM_PWM_ConfigChannel(&htim2, &sConfigOC, TIM_CHANNEL_2);

}

void MX_GPIO_Init(void)
{

    GPIO_InitTypeDef GPIO_InitStruct;
```

```c
  __GPIOC_CLK_ENABLE();
  __GPIOF_CLK_ENABLE();
  __GPIOA_CLK_ENABLE();
  __GPIOB_CLK_ENABLE();

  GPIO_InitStruct.Pin = GPIO_PIN_13;
  GPIO_InitStruct.Mode = GPIO_MODE_EVT_RISING;
  GPIO_InitStruct.Pull = GPIO_NOPULL;
  HAL_GPIO_Init(GPIOC, &GPIO_InitStruct);

  GPIO_InitStruct.Pin = GPIO_PIN_2|GPIO_PIN_3;
  GPIO_InitStruct.Mode = GPIO_MODE_AF_PP;
  GPIO_InitStruct.Pull = GPIO_NOPULL;
  GPIO_InitStruct.Speed = GPIO_SPEED_LOW;
  GPIO_InitStruct.Alternate = GPIO_AF7_USART2;
  HAL_GPIO_Init(GPIOA, &GPIO_InitStruct);

  GPIO_InitStruct.Pin = GPIO_PIN_5;
  GPIO_InitStruct.Mode = GPIO_MODE_OUTPUT_PP;
  GPIO_InitStruct.Pull = GPIO_NOPULL;
  GPIO_InitStruct.Speed = GPIO_SPEED_LOW;
  HAL_GPIO_Init(GPIOA, &GPIO_InitStruct);

}

void HAL_TIM_PeriodElapsedCallback(TIM_HandleTypeDef *htim)
{

  HAL_ADC_Start(&hadc1);
  HAL_ADC_PollForConversion(&hadc1, 10);
  data=HAL_ADC_GetValue(&hadc1);
  cont++;

  if (data>1861 && flag==0) {
    nPeak++;
    flag=1;
  }
  if (data<1861 && flag==1) {
    flag=0;
  }
  if (cont==1000) {
    nP2=nPeak;
    if (nPeak<5) {
      TIM2->CCR1=500;
      TIM2->CCR2=0;
    } else {
      TIM2->CCR1=0;
      TIM2->CCR2=50;
    }
    cont=0;
    nPeak=0;
  }
}

#ifdef USE_FULL_ASSERT
void assert_failed(uint8_t* file, uint32_t line)
{
}
#endif
```

File **stm32f3xx_hal_msp.c**

```c
#include "stm32f3xx_hal.h"

void HAL_MspInit(void)
{
  HAL_NVIC_SetPriorityGrouping(NVIC_PRIORITYGROUP_0);

  HAL_NVIC_SetPriority(SysTick_IRQn, 0, 0);

}

void HAL_ADC_MspInit(ADC_HandleTypeDef* hadc)
{

  GPIO_InitTypeDef GPIO_InitStruct;
  if(hadc->Instance==ADC1)
  {
    __ADC12_CLK_ENABLE();

    GPIO_InitStruct.Pin = GPIO_PIN_0;
    GPIO_InitStruct.Mode = GPIO_MODE_ANALOG;
    GPIO_InitStruct.Pull = GPIO_NOPULL;
    HAL_GPIO_Init(GPIOA, &GPIO_InitStruct);

  }

}

void HAL_ADC_MspDeInit(ADC_HandleTypeDef* hadc)
{

  if(hadc->Instance==ADC1)
  {
    __ADC12_CLK_DISABLE();

    HAL_GPIO_DeInit(GPIOA, GPIO_PIN_0);

  }

}

void HAL_TIM_Base_MspInit(TIM_HandleTypeDef* htim_base)
{

  GPIO_InitTypeDef GPIO_InitStruct;
  if(htim_base->Instance==TIM1)
  {
    __TIM1_CLK_ENABLE();
    HAL_NVIC_SetPriority(TIM1_UP_TIM16_IRQn, 0, 0);
    HAL_NVIC_EnableIRQ(TIM1_UP_TIM16_IRQn);
  }
  else if(htim_base->Instance==TIM2)
  {
    __TIM2_CLK_ENABLE();

    GPIO_InitStruct.Pin = GPIO_PIN_1|GPIO_PIN_15;
    GPIO_InitStruct.Mode = GPIO_MODE_AF_PP;
    GPIO_InitStruct.Pull = GPIO_NOPULL;
    GPIO_InitStruct.Speed = GPIO_SPEED_LOW;
    GPIO_InitStruct.Alternate = GPIO_AF1_TIM2;
    HAL_GPIO_Init(GPIOA, &GPIO_InitStruct);
  }
```

```
}

void HAL_TIM_PWM_MspInit(TIM_HandleTypeDef* htim_pwm)
{

  GPIO_InitTypeDef GPIO_InitStruct;
  if(htim_pwm->Instance==TIM2)
  {
    __TIM2_CLK_ENABLE();

    GPIO_InitStruct.Pin = GPIO_PIN_1|GPIO_PIN_15;
    GPIO_InitStruct.Mode = GPIO_MODE_AF_PP;
    GPIO_InitStruct.Pull = GPIO_NOPULL;
    GPIO_InitStruct.Speed = GPIO_SPEED_LOW;
    GPIO_InitStruct.Alternate = GPIO_AF1_TIM2;
    HAL_GPIO_Init(GPIOA, &GPIO_InitStruct);

  }

}

void HAL_TIM_Base_MspDeInit(TIM_HandleTypeDef* htim_base)
{

  if(htim_base->Instance==TIM1)
  {
    __TIM1_CLK_DISABLE();

    HAL_NVIC_DisableIRQ(TIM1_UP_TIM16_IRQn);

  }
  else if(htim_base->Instance==TIM2)
  {
    __TIM2_CLK_DISABLE();

    HAL_GPIO_DeInit(GPIOA, GPIO_PIN_1|GPIO_PIN_15);

  }

}

void HAL_TIM_PWM_MspDeInit(TIM_HandleTypeDef* htim_pwm)
{

  if(htim_pwm->Instance==TIM2)
  {
    __TIM2_CLK_DISABLE();

    HAL_GPIO_DeInit(GPIOA, GPIO_PIN_1|GPIO_PIN_15);

  }

}
```

Bibliography

Abarbanel, H. (2012). *Analysis of observed chaotic data* (Springer).

Andò, B. and Graziani, S. (2012). *Stochastic resonance: theory and applications* (Springer).

Awrejcewicz, J. and Koruba, Z. (2012). *Classical mechanics. Applied mechanics and mechatronics* (Springer).

Baranyi, A. and Chua, L. O. (1982). Dynamic model for the analog multiplier, *IEEE Transactions on Circuits and Systems* **29**, 2, pp. 65–76.

Bendat, J. S. and Piersol, A. G. (2011). *Random data: analysis and measurement procedures* (Wiley).

Belato, D., Weber, H. I., Balthazar, J. M. and Mook, D. T. (2001). Chaotic vibrations of a non-ideal electro-mechanical system, *International Journal of Solids and Structures*, **38**, 10, pp. 1699–1706.

Berg, J. M. and Wickramasinghe, I. M. (2015). Vibrational control without averaging, *Automatica* **58**, pp. 72–81.

Bfazejezyk, B., Kapitaniak, T., Wojewoda, J. and Brindley, J. (1993). Controlling chaos in mechanical systems, *Appl Mech Rev*, **46**, 7, pp. 385–391.

Bhattacharyya, S. P., Chapellat, H. and Keel, L. H. (1995). *Robust control: the parametric approach* (Upper Saddle River).

Bonato Altran, A., Chavarette, F. R., Minussi, C. R., Peruzzi, N. J., Marthins Lopes, M. L., Barbanti, L. and Damasceno, B. C. (2011). Control design applied to a micro electro mechanical system: MEMS Comb Drive, *Advanced Materials Research* **217-218**, pp. 33–38.

Braiman, Y., Lindner, J. F. and Ditto, W. L. (1995). Taming spatiotemporal chaos with disorder, *Nature*, **378**, 6556, pp. 465–467.

Bucolo, M., Caponetto, R., Fortuna, L., Frasca, M. and Rizzo, A. (2002). Does chaos work better than noise? *IEEE Circuits and Systems Magazine*, **2**, 3, pp. 4–19.

Buscarino, A., Fortuna, L., Frasca, M., and Muscato, G. (2007). Chaos does help motion control, *International Journal of Bifurcation and Chaos* **17**, 10, pp. 3577–3581.

Buscarino, A., Fortuna, L., Frasca, M. and Gambuzza, L. V. (2012). A chaotic circuit based on Hewlett-Packard memristor, *Chaos*, **22**, 2, 023136.

Buscarino, A., Fortuna, L., Frasca, M. and Gambuzza, L. V. (2013). A gallery of chaotic oscillators based on HP memristor, *International Journal of Bifurcation and Chaos*, **23**, 5, 1330015.

Buscarino, A., Fortuna, L., Frasca, M. and Sciuto, G. (2014). *A concise guide to chaotic electronic circuits*, (Springer).

Buscarino, A., Corradino, C., Fortuna, L., Frasca, M. and Sprott, J. C. (2016). Nonideal behavior of analog multipliers for chaos generation, *IEEE Transactions on Circuits and Systems II: Express Briefs* **63**, 4, pp. 396–400.

Buscarino, A., Famoso, C., Fortuna, L. and Frasca, M. (2016). Passive and active vibrations allow self-organization in large-scale electromechanical systems, *International Journal of Bifurcation and Chaos* **26**, 7, pp. 1650123.

Buscarino, A., Famoso, C., Fortuna, L. and Frasca, M. (2016). A new chaotic electro-mechanical oscillator, *International Journal of Bifurcation and Chaos* **26**, 10, pp. 1650161.

Buscarino, A., Fortuna, L. and Frasca, M. (2010). *Essentials of nonlinear circuit dynamics with MATLAB and laboratory experiments* (CRC Press).

Busch-Vishniac, I. J. (2012). *Electromechanical sensors and actuators* (Springer).

Chen, G. and Yu, X. (2003). *Chaos control: theory and applications*, (Springer Science & Business Media).

Chotorlishvili, L., Ugulava, A., Mchedlishvili, G., Komnik, A., Wimberger, S. and Berakdar, J. (2011). Nonlinear dynamics of two coupled nano-electromechanical resonators, *Journal of Physics B: Atomic, Molecular and Optical Physics*, **44**, 215402.

Cook, P. A. (1194). *Nonlinear dynamical systems* (Prentice Hall).

Duffing, G. (1918). *Erzwungene schwingungen bei vernderlicher eigenfrequenz und ihre technische bedeutung*, (R, Vieweg & Sohn).

Erturk, A. and Inman, D. J. (2011). Broadband piezoelectric power generation on high-energy orbits of the bistable Duffing oscillator with electromechanical coupling, *Journal of Sound and Vibration* **330**, pp. 2339–2353.

Feodosev, V. I. (1977). *Rsistance des matriaux: problems et questions choisis* (Mir).

Fortuna, L., Frasca, M. and Rizzo, A. (2003). Chaotic pulse position modulation to improve the efficiency of sonar sensors, *IEEE Transactions on Instrumentation and Measurement*, **52**, 6, pp. 1809–1814.

Fortuna, L., Frasca, M., Graziani, S. and Reddiconto, S. (2006). A chaotic circuit with ferroelectric nonlinearity, *Nonlinear Dynamics*, **44**, 1, pp. 55–61.

Fortuna, L., Frasca, M. and Xibilia, M. G. (2009). *Chua's circuit implementations: yesterday, today and tomorrow* (World Scientific).

Garcia-Ojalvo, J. and Sancho, J. M. (1999). *Noise in spatially extended systems* (Springer).

Gawronski, W. (2006). *Balanced control of flexible structures* (Springer).

Harada, Y., Masuda, K. and Ogawa, A. (1996). Dynamical behavior of acoustically coupled chaos oscillators, *Fractals*, **4**, pp. 407–414.

Holmes, P. (1979). A nonlinear oscillator with a strange attractor, *Philosophical Transaction Royal Society A*, **292**, 1394, pp. 419–448.

Hurmuzlu, Y and Nwokah, O. D. I. (2001). *The mechanical systems design hand-*

book: modeling, measurement, and control (CRC Press).

Illing, L., Fordyce, R. F., Saunders, A. M. and Ormond, R. (2012). Experiments with a MalkusLorenz water wheel: chaos and synchronization, *American Journal of Physics* **80**, 3, pp. 192–202.

Itoh, M. and Chua, L. O. (2008). Memristor oscillators, *International Journal of Bifurcation and Chaos*, **18**, 11, pp. 3183–3206.

Johnson, M. A. and Moon, F. C. (1999). Experimental characterization of quasiperiodicity and chaos in a mechanical system with delay, *International journal of Bifurcation and Chaos* **9**, 1, pp. 49–65.

Kaplan, H. (2007). *Practical applications of infrared thermal sensing and imaging equipment* (SPIE press).

Karnopp, D. (1985). Computer simulation of stick-slip friction in mechanical dynamic systems, *Transactions of the ASME Journal of Dynamic Systems, Measurement and Control* **107**, pp. 100-103.

Kolár, M. and Gumbs, G. (1992). Theory for the experimental observation of chaos in a rotating waterwheel, *Physical review A* **45**, 2, pp. 626-637.

Letellier, C. (2013). *Chaos in nature*, (World Scientific).

Levine, W. S. (1996). *The control handbook*, (CRC Press).

Lichtenberg, A. J. and Lieberman, M. A. (2013). *Regular and stochastic motion*, (Springer Science & Business Media).

Madan, R. N. (ed.) (1993). *Chua's circuit: a paradigm for chaos* (World Scientific).

Meerkov, S. M. (1973). Vibrational control, *Avtomat. i Telemekh*, 2, pp. 34–43.

Meerkov, S. M., Kabamba, P. T. and Poh, E. K. (1993). *Closed loop vibrational control: theory and applications* (U.S. Army Research Office).

Mohan, N., Undeland, T. M. and Robbins, W. P. (1988). *Power electronics* (Wiley).

Moon, F. C. (1980). Experiments on chaotic motions of a forced nonlinear oscillator: strange attractors, *Journal of Applied Mechanics* **47**, 3, pp. 638–644.

Moon, F. C. (2004). *Chaotic vibrations*, (Wiley).

Moskowitz, L. R. (1995). *Permanent magnet design and application handbook* (Krieger).

Muthuswamy, B. (2010). Implementing memristor based chaotic circuits, *International Journal of Bifurcation and Chaos* **20**, 5, pp. 1335–1350.

Nijmeijer, H. and Rodriguez-Angeles, A. (2003). *Synchronization of mechanical systems* (World Scientific).

Ott, E. (2002). *Chaos in dynamical systems*, (Cambridge University Press).

Owens, B. A. M., Stahl, M. T., Corron, N. J., Blakely J. N. and Illing, L. (2013). Exactly solvable chaos in an electromechanical oscillator, *Chaos* **23**, 033109.

Priemer, R. (1991). *Introductory signal processing* (World Scientific).

Rosenstein, M. T., Collins, J. J. and De Luca, C. J. (1993). A practical method for calculating largest Lyapunov exponents from small data sets, *Physica D* **65**, pp. 117–134.

Shaw, S. W. and Rand, R. H. (1989). The transition to chaos in a simple mechanical system, *International Journal of Non-Linear Mechanics*, **24**, 1, pp. 41–56.

Siewe Siewe, M., Yamgoué, S. B., Moukam Kakmeni, F. M. and Tchawoua C. (2010). Chaos controlling self-sustained electromechanical seismograph system based on the Melnikov theory, *Nonlinear Dynamics*, **62**, pp. 379–289.

Skvarenina, T. L. (ed.). (2001). *The power electronics handbook* (CRC press).

Slotine, J. J. E. and Li, W. (1991). *Applied nonlinear control* (Prentice-Hall).

Sprott, J. C. (2003). *Chaos and time-series analysis*, (Oxford University Press).

Sprott, J. C. (2010). *Elegant chaos: algebraically simple chaotic flows* (World Scientific).

Steur, E. (2007). *On synchronization of electromechanical Hindmarsh-Rose oscillators*, (Eindhoven University of Technology).

ST Microelectronics (2014). \http://www.st.com/web/en/catalog/mmc/FM141/SC1169/SS1576?sc=stm32f3

Sun, J.-Q. (2006). *Stochastic dynamics and control* (Elsevier).

Takens, F. (1981). Detecting strange attractors in turbulence, in *Dynamical systems and turbulence*, (Springer).

Thomsen, J. J. (2003). *Vibrations and stability* (Springer).

Ueda, Y. (1979). Randomly transitional phenomena in the system governed by Duffing's equation, *Journal of Statistical Physics* **20**, pp. 181–186.

Webster, J. G. and Eren, H. (eds.) (2014). *Measurement, instrumentation, and sensors handbook: electromagnetic, optical, radiation, chemical, and biomedical measurement* (CRC press).

Yim, G. S., Ryu, J. W., Park, Y. J., Rim, S., Lee, S. Y., Kye, W. H. and Kim, C. (2004). Chaotic behaviors of operational amplifiers, *Physical Review E* **69**, 4, pp. 045201.

Index

www.ingramcontent.com/pod-product-compliance
Lightning Source LLC
Chambersburg PA
CBHW050631190326
41458CB00008B/2223